SpringerWienNewYork

Studies in Space Policy
Volume 6

Edited by the European Space Policy Institute
Director: Kai-Uwe Schrogl

Jacques Arnould

Icarus' Second Chance

The Basis and Perspectives of Space Ethics

SpringerWienNewYork

Dr. Jacques Arnould

Centre National d'Etudes Spatiales, Paris, France

This work contains an extended and revised translation from the French language original "La Seconde Chance d'Icare" by Jacques Arnould, published 2001 by Les Éditions du Cerf, additionally including translated material from the work "La pollution spatiale sous surveillance", published 2007 by Edition Ellipses.

SpringerWienNewYork is a part of
Springer Science + Business Media
springer.at

Typesetting: Thomson Press (India) Ltd., Chennai
Printing: Strauss GmbH, 69509 Mörlenbach, Germany

Cover: Stefan Seip – astromeeting.de.

Printed on acid-free and chlorine-free bleached paper

SPIN: 80035158

With 8 Figures

Library of Congress Control Number: 2011931322

ISSN 1866-5307

ISBN 978-3-7091-0711-9 SpringerWienNewYork

Preface

Whatever part we have played in the incredible programmes and organisations that have enabled space to be explored and conquered, all of us must have wondered at some point about the ethics of our work and our commitment, asking ourselves whether or not it would prove to be of benefit to mankind or would lead to progress for humanity.

Late July 1969, three of us circled the Moon on board the Columbia command module. Then, Neil Armstrong and I received the go ahead to land the Eagle lunar module on the Moon's surface. Our minds were completely focused on the many tasks to be performed and on the success of our dangerous descent: there wasn't time to spare for asking too many questions about the ethical nature of our mission! Having landed, I very quickly weighed up the impact of our Moon landing on the future of mankind. Neil summed it up in his own way by uttering the famous words that will always be associated with this adventure. I personally chose to mark the occasion in silence, before opening the LEM airlock, and by thanking God, asking him to protect mankind as it embarked on a new era.

I therefore view Jacques Arnould's ethical study of space venture as both pertinent and remarkable. It deals with a number of aspects and raises a number of questions, not only about the presence of humans in space, but also about the development of communication, remote sensing and geopositioning applications; everything, in short, that contributes to the globalisation of our society on Earth. Such progress obviously has moral and social consequences.

Space offers us, or rather has allowed us to adopt for ourselves, a new dimension of freedom, which we must use for the benefit of humanity, to enrich and not degrade our lives. The notions presented in this book may well spark a debate that I personally hope is as rich and global as possible.

Buzz Aldrin, Astronaut, Apollo XI

Table of contents

"Is there anything nobler than becoming master of all the Sun's energy,
two billion times greater than the energy that reaches the Earth?!
Is there anything more beautiful than escaping our humble planet,
to enter the greater cosmos and to offer mankind the opportunity
to move beyond the tight constraints of life on Earth
and to break free from the chains of gravity?!"
Konstantin Tsiolkovsky, *Komsomolskaïa Pravda*, July 23rd, 1935.

"We set sail on this new sea because there is new knowledge to be gained,
and new rights to be won, and they must be won and used for the progress
of all people. For space science, like nuclear science and all technology,
has no conscience of its own."
John F. Kennedy, September 12th, 1962.

Introduction

2011: half a century has now passed since Yuri Gagarin's historic flight. Icarus' dream was finally fulfilled. Mankind, after exploding the confinements of the crystal spheres of ancient astronomy and understanding that the Earth, its Earth, was not at the centre of the cosmos but was just a speck among billions of stars and other celestial bodies, finally managed to break free from the irresistible pull of gravity that had held it and all the bodies surrounding it, so firmly to its native planet. Through the ingenuity of Konstantin Tsiolkovsky, Robert Goddard and Robert Esnault-Pelterie, their engineering colleagues and their successors, through the curiosity and tenacity of scientists, the political and financial commitment of the first nations in space and now of a growing number of countries and companies, through the courage of astronauts, cosmonauts and all men and women who are still made of "the right stuff", and finally through the support, differing in many ways, of public opinion; as a result of all of the above, and also through what a historian would sometimes refer to as chance, luck or opportunism, mankind broke through Earth's atmosphere, that frontier enveloping our planet, and began to send man-made machines and humans into space.

Therefore, 50 years on from Gagarin and only a little longer since the launch of Sputnik 1, each of us will interpret and weigh up in our own way this period of time during which humanity has lived through so many significant events, has suffered and caused so many disasters and has made so much progress to which space has very often contributed, benefited from or been a party to. I am convinced that space is profoundly human, even humanist and I will expand on this in the book.

How, therefore, could space activities not give rise to ethical questions or provoke ethical thought? I know that people may have an aversion to the word "ethics" and that it can often put them off but we need to come to terms with it, as this book is specifically devoted to the ethics of space. Space ethics: the term describes a field of activity and research which humans have not yet fully considered or which has only been dealt with sporadically and yet which is developing to offer an original vision of space agencies, their practices and their resources, their policies and their projects. I am convinced that ethics truly constitutes a new frontier for the great space adventure.

CHAPTER 1

1 Is the sky open to us?

In 1960, three years after the launch of the first *sputnik* by the Soviets and one year before Yuri Gagarin's flight, the German Walter Pons published a work entitled: *Steht uns der Himmel offen?* (Is the sky open to us?).[1] This work is probably one of the first philosophical studies conducted on the astronautics venture, once this actually became reality. To the question which forms the title of this book, its author replies: "We cannot really know the world if we do not firstly know ourselves". This answer is suggestive of Socrates as the famous Greek philosopher transformed the motto engraved on the front of the temple at Delphi "Know Thyself and leave the World to the Gods", into: "Know thyself and you will know the Universe and the Gods". In this second version, Hegel identified a major turning point for humankind: in effect, Socrates was proposing to make the inner conscience the authority of truth and therefore of decision. It was no longer necessary, or even an option, to leave things to be sorted out by a superior and unattainable divine order, that could only be revealed by the oracle, by divination and mysticism. It was henceforth up to man to take his destiny in his own hands.

So be it. Yet why associate knowledge of the self with that of the universe in this way, when even the priests and the oracles of Delphi encouraged their loyal followers to ignore the universe (my omission of the question of gods is deliberate)? Should we recognise in the universe, and here I am referring to the sky as defined by Pons, a unique relationship with the knowledge which man may have of himself? Is it possible that the sky, the universe or what I will refer to here as space, in the same way as the tree of knowledge of good and evil growing in the mythical garden of Eden, could be forbidden to humans through fear they might take themselves for gods or become like gods, or through fear that they might discover their true nature, their powers and their limits? If such is the case, it is easy to understand Pons' question: is the sky open to us? Is it possible or even reasonable to aspire to reach it? What might we learn about ourselves?

1.1 What is space?

Before embarking on any philosophical and ethical discussions, we must first ask ourselves what space really is and define the meaning of the term. In its *Dictionnaire de spatiologie* (Dictionary of Aeronautics and Space Technology), the Conseil international pour la langue française (International Council for the French Language) gives the following definition: "1. Common shortened form of outer space. 2. Field of human activities pertaining to outer space". What should we understand by outer space? The "region of the Universe located beyond the part of the Earth's atmosphere in which aircraft can manoeuvre. The expression 'outer space' is used in space law without specific definition or delimitation. The lower limit of outer space cannot be associated with a specific altitude but it is generally given that it is around 50 km. However, spacecraft can experience deceleration or a rise in temperature caused by the atmosphere, at much higher altitudes".[2]

"Without specific definition or delimitation": the official definition in this dictionary may surprise you, but it is perfectly correct. Neither the lower or upper limit of space can be currently given or specified. They can only be estimated. In effect, while it may be possible to set a lower limit for space, often an altitude of 100 km, the upper limit remains unknown and is the subject of much scientific speculation: is the universe finite, infinite or unlimited? So many reasons for abandoning our ancestors' belief in "solid" space or solid ether, or any other definition of space that is too physical or too geographical, in favour of a more technical approach: space is first and foremost a set of moving bodies, not only natural celestial bodies but now also manmade devices, even human beings, a set of trajectories and movements, knowledge and techniques, communications and relationships. Space is not only a place, it is also what humans do there and what they make of it. Its limits are not only those imposed by its geography or its natural characteristics but also those resulting from our knowledge and ignorance alike, with regard to science and technology.

The situation is interesting to say the least. On the one hand, our imagination is constantly stimulated by the enhancement of scientific knowledge and technological expertise in matters of space, continually providing us with new sources of inspiration whilst, on the other hand, using exploration and observation techniques, space is now a more tangible reality and closer to us than the spheres, orbs, epicycles or even multiple universes imagined by our ancestors. This overlap between the fruits of the imagination, going back centuries or even millennia, and the new and constantly renewed components making up reality, provokes both astonishment and excitement, but also frustration and disappointment, yet rarely indifference. Space is never totally alien to humans.

However, it is important to talk about one particular sticking point: the status of the Earth. Excluded from the definition of space, here qualified as extra-atmospheric which is synonymous with extra-terrestrial, our Earth still very much belongs to what our ancestors called, and what we continue to refer to today as "the cosmos". It is just one of the planets that make up the solar system, one of hundreds of planets known to and observed by astronomers. How is it possible to account for this common classification, whilst respecting the uniqueness, clearly marked by anthropocentrism, of our planet? What relationship can we envisage and establish between Earth and space, one which is not only technical but also legal, cultural and finally ethical? Could we view space as some kind of suburb of Earth, a suburb with undefined boundaries? Conversely, should the Earth be reduced to the status of a speck, lost within what appears to be first and foremost an "indifferent immensity" (Jacques Monod)? What can we say, in this context, of mankind's astronautical achievements? Should we refer to giant leaps or pretentious, ridiculous flea jumps? When such questions are asked, ethics is just around the corner.

1.2 What is ethics?

Ethics or the task of being human. It is not easy to give a satisfactory definition of "ethics", probably because the term is one of those words whose usage is as widespread as its meaning is approximate or diversified. Various philosophical, political or religious conventions thus suggest an art of living and of dying (*modus vivendi* and *modus moriendi*), to define a purpose (Greek *telos*) which can often only be achieved after a whole lifetime of experiences, practical customs and the fruit of repetition. Alternatively, they place emphasis on the one-off act, which must be referred to a law, to a necessity, to a standard, or to an internal or external obligation. It would therefore seem that no definition of ethics is satisfactory. To the most common ("Do good, not evil", "Choose to become ever more humane") and to that of French scientist Jean Bernard ("ethics guarantees the harmony which results from correct behaviour of the soul to ensure that all things – and all acts – in the world are in their rightful place"[3]), I add that of Michel Foucault: "By attitude, I mean a method of relating to reality; a voluntary choice made by some; that is, a way of thinking and feeling, also a way of acting and behaving which, all together, marks a belonging and presents itself as a task. Probably similar to that which the Greeks called an *ethos*)".[4]

1.2.1 Ethics and morality, deontology and law

Modern usage of the term "ethics" involves a reflexive and philosophical meaning and morality is mostly used to designate rules of conduct, especially if they have an imperative or collective character (possibly giving this second term a pejorative note). The main importance of this distinction is to remind us that all rules, standards or laws claiming to govern our individual or collective behaviour can and should be subject to an analysis in terms of actions and free will, value and purpose: where does free will come in, in relation to action, based on a value and relative to a purpose? What about Goodness, Truth, Beauty? Who is the subject, the moral agent who says "I"? The works of American authors sometimes employ the abbreviation EVS, which stands for *Ethics and Values Studies*. I believe that this properly takes into account the importance to be placed on the philosophical dimension of ethical reflection. In effect, if ethics and morals are permanently confronted with the concrete and practical nature of human existence and the reality of things, if they express and are materialised most often in terms of obligations, musts or needs, the subject must be able to justify his/her judgement, to follow his/her conscience according to criteria other than simple blind obedience to a law which, from a freely accepted standard, may still become a restrictive law. In other words, ethics and morals must provide individuals with a framework inside which they can be exposed to others and interact with mutual respect. This is the meaning of the definition given by Foucault and all those who associate morals with actions, thought with risk and recognition with the other.

We must not therefore focus entirely on questioning ourselves about the adequacy of means to ends (to which utilitarian ethics is most often reduced), on finding a *modus vivendi*, a way of living decently or even, from a pessimistic point of view, a *modus morendi*, a way of dying. It is also important to think philosophically and sociologically about the way in which people and organisations define their values and their purposes, the way in which they implement them, negotiate them with their partners and consider future generations. This approach must continually seek to clarify standards, values, membership systems, often inherited from past experience and collective memory, to consider the complexity and singularity of real situations. In fact, in most situations, there is not one single and definitive answer to our questions but a variety of answers from which we need to choose. Engineers experience this whenever they design and develop a project.

So, ethics is (or should be) more than a trend, a justification, a "witch hunt" or an opportunity to instil fear. On the contrary, ethics is an essential characteristic of the human species, made up of conscious, intelligent and free beings driven by values and beliefs, confronted at the same time, just like other living beings, with the need to ensure their survival, their preservation and their future. Humans constantly wonder about the content, meaning and consequences of their actions and those of others, regardless of who is the instigator, the victim or the onlooker. Judgements and assessments, decision-making and choices and confrontation between what is commonly designated as nature and nurture are all areas in which an individual takes an ethical stance, often without even knowing it. Individuals just need the means to apply and develop this stance ever more consciously and with increasing awareness.

Even though, etymologically speaking, ethics, morals and deontology are practically synonymous, the latter was historically associated with professional experience and practice (medicine, law, notary public, architecture, etc.). The term therefore designates duties associated with performing a profession, combined in the form of rules and codes. Guy Durand explains that "Deontology codes often contain, in addition to veritable ethical and moral standards, administrative rules intended to ensure quality of professional work and protect the reputation of the corporation. On the other hand, their existence supposes that these rules have been formally adopted by a professional or political authority, that they are subject to a certain consensus of members and that they do not require too much heroism. This explains a certain convergence of deontology codes with the notion of law (positive law) and a concomitant distancing in relation to the notion of ethics and morals)".[5]

What places does law have in ethics? Probably somewhere between morals and deontology. Along with Durand, I share the opinion according to which ethics and morals can require or lead to heroic acts; ethics pertains to the search for and adherence to principles and values: it focuses on the obligations of the inner self. Law, which is fundamentally based on ethics, represents exteriority. It implies compliance with an exterior and minimum rule (e.g. talion law: "An eye for an eye, a tooth for a tooth") and supposes the legitimacy of coercion and finally requires an attitude of submission from individuals, without any other form of questioning. Ethics and morality can lay claim to a form of universality. For its part, and out of necessity, law is more modest, limited to specific groups and eras (with rare exceptions or attempts such as the *Declaration of the Rights of Man*) and is in this way more susceptible to being modernised, improved and questioned.

1.2.2 Codes of ethics and ethics committees

What is an ethics committee? It is primarily defined by its objective which is to establish recommendations, aid public opinion, political and administrative leaders and standardisation authorities. To fulfil this task, members of such a committee must prove themselves capable of objectivity, sound judgement as well as being independent and competent. They must possess a wealth of qualities and knowledge, affiliations and inherent personality traits. The interdisciplinarity of a committee is indispensable. The committee may be required to give opinions, ensure adherence to deontology, a code of conduct, or even a given idea of morality and ethics. Its opinions may also be used by society's competent authorities to draw up legislative and juridical standards. If the approach adopted is primarily descriptive, if it focuses on the sociocultural context, which incorporates the problem being discussed, it is set between two trains of logical thought, one operational and the other the "sense of existence".

An ethics committee also simultaneously plays the role of arbiter and go-between. Gérard Toulouse stated "It is desirable that all scientists or all citizens may notify an Ethics Committee and that their concern be, if necessary, passed from Committee to Committee at different levels". He adds: "These Committees must find a middle way, between two clear pitfalls. The first is excessive 'ethical correctness' [. . .]. The second pitfall – more threatening in France – is confinement, i.e. the restriction of ethics to a limited domain, and its confiscation by experts".[6] The reader will have understood, the subject of these pages clearly does not support a limitation or seizure of ethics by a few specialists and experts.

Whilst ethics committees appeared in the 1980s, codes of ethics date back much further than the point at which our societies started questioning technologies and sciences. On the one hand, these codes feature in the deontological perspective of vows, well known in the healthcare professions, and on the other hand, they are one of the expressions of a positive awareness of values and responsibilities associated with a profession. This is particularly true of engineers. Being an engineer is not simply about having a mere job, it is about practising a profession, performing a vocation, with its requirement for honesty, expertise and professionalism, respect for others, their health and their rights, universality, autonomy and confidentiality. Engineers must also have a sense of community and society, and advocate progress for humanity along with respect for the environment. This is probably one of the central ideas of codes of ethics, as the National Society of Professional Engineers (NSPE) reminds us in its preamble: "Engineering is an important and learned profession. As members of this profession, engineers are expected to exhibit the highest standards of honesty and integrity. Engineering has a direct and vital impact on the quality of life for all people. Accordingly, the services provided by engineers require honesty, impartiality, fairness, and equity, and must be dedicated to the protection of the public health, safety, and welfare. Engineers must perform under a standard of professional behavior that requires adherence to the highest principles of ethical conduct".[7] The general framework of this code, as with similar codes, is therefore that of modern Western thought, marked by humanism and the concept of "human honesty" and by the role granted to progress. Engineers (but I suppose they are not alone here) must effectively manage two types of relationship: a relationship with the professional body to which they are affiliated and a relationship with the society in which and for which they work. Among the qualities engineers need to develop to meet this requirement, expertise occupies a central place in the NSPE's code: engineers must always maintain a very high level of theoretical knowledge and practical skill, without, however, overstepping boundaries or letting things go to their heads. However, in reading this code, exercising skill would seem to be a difficult balance to find and maintain since it does not depend solely on the engineer, but often on the work of other professionals, based on coordination and a relationship of responsibility that must always be properly defined. Associated with skill is responsibility, in terms of which, it should be noted that, since engineers in theory freely adhere to a code, there may not be repression or punishment outside of the common regulatory framework. At the very most it may express obligations, as the NSPE code does. For example, the prohibition of taking part in coercive actions (e.g. a strike), the engineer's undertaking to respect the confidential nature of information relating to the economic affairs and technical processes of his/her employer, the rejection of the compromise, in its various guises, of principles and commission, the respect of intellectual property and that of competition, and so on.

All things considered, what purpose do codes of ethics serve? Are they sticks or carrots? Do they not help to make the professions which they define sacred? Was this the risk that George Bernard Shaw was denouncing when he stated: "All professions are conspiracies against the laity"? The first engineering codes appeared at the start of the 20th century, but we need to recognise that their effectiveness has always remained limited, probably because this environment, unlike that of the health professions, is subject to major contractual and wage constraints. This is probably the reason some engineering associations have created awards to reward those of their members who have, sometimes at the risk of their job, complied with the code and values of their profession! In any case, a code serves to specify, explain and even remind us of the purposes, skills and responsibilities of a professional group. In this perspective, it is as much a carrot as a stick.

1.2.3 Putting experts to good use

Experts have today become an essential reference for our societies (and not only in television series) be they involved in taking actions or estimating the consequences of actions taken. It is doubtlessly for this reason that expertise is, at the same time, called into question. The knowledge and skills of theses expert men and women, or those responsible for the expertise, are not generally very difficult to assess. Depending on the circle to which they belong and their affiliations, they are subject to deontology and rules of operation, to publishing regulations and, if they are scientists, to the confrontation with common knowledge. In short, skill and authority are not about asserting one's opinion or imposing it by force. The challenge inherent in expertise would appear to principally involve how to manage, even slow down, the slide which can tend to occur between knowledge and power. Former President of the French National Centre for Scientific Research (CNRS) Gérard Mégie wrote that scientists are responsible for "managing knowledge as part of their role as scientific researcher, acting as an expert knowing that even if they do not necessarily have the exact answer, they must give an opinion and use it to reach beyond the limits of their own knowledge". Exceeding the limits of one's knowledge and skills – this naturally involves a temptation to which scientists are not the only ones to be subjected. This temptation should not be underestimated or disregarded.

In his work *Entre savoir et décision, l'expertise scientifique* (Between knowledge and decision, scientific expertise), Philippe Roqueplo highlights and analyses this possible slide from science to expertise; as the title of his book suggests, he is questioning what the appropriate role should be for the expert, on the one hand, and expertise on the other: between knowledge and decision.

The term "expert" itself, he notes, may cause ambiguity, depending on whether we are using the noun or the adjective. With no specific indication, the noun effectively designates a person whose role it is to draft expert reports. The qualifier is synonymous with skilled, qualified: you can have confidence in the person who possesses or who has deserved the description as they "know", they know what the question is. They may thus be conferred, entrusted with a responsibility, a task or even the status of negotiator or political player. The possible confusion then becomes evident: the "expert" provides knowledge that may allow the person deemed to be "expert" to make a choice or a decision, to act. While the latter does not need the former, the former cannot take the place of the latter, unless claiming a power which is not theoretically his/her own. Correctly managing this difference, which transcends the correct usage of the noun and the adjective, pertains to deontology and ethics.

So, Philippe Roqueplo speaks in favour of the institutionalisation of permanent expertise preparation procedures: "This appears to me to demand that, under the supervision of the ministries on which they depend, scientific organisations implement (intra and inter organisation) expertise procedures intended not to reach a consensus but, on the contrary, to widen, as far as possible, the gap of the scientific criticism of conceivable options: it is this gap that ensures there is enough room to make decisions".[8]

Is the sky open to us? Engineers specialising in astronautical technologies and space programme managers can and must provide different answers to this question. They must in particular distinguish between performance (will the implemented technology work?) and design (is the implemented technology adapted to the requirement, to its maintenance?). Formulating these requirements is sometimes as difficult as providing the technical solutions, if not more so, up to the point of slowing down the progress of knowledge and expertise. Space does not escape this fact: as its history reveals, it depends as much on political decisions and public support as on technical discoveries. Think, decide, act: there was no way that space could not escape ethical questions.

[1] Pons, Walter. *Steht uns der Himmel offen? Entropie-Ektropie-Ethik. Ein Beitrag zur Philosophie des Weltraumzeitalters.* Wiesbaden: Krausskopf Verlag, 1960.
[2] French Space Agency CNES – International Council for the French Language. *Dictionnaire de Spatiologie. Sciences et techniques spatiales*, vol. 1: *Termes et définitions.* Paris, 2001: 101.
[3] Quoted in Pompidou, Alain. *Souviens-toi de l'homme. L'éthique, la vie, la mort.* Paris: Payot, 1990: 10.
[4] Foucault, Michel. "Qu'est-ce que les Lumières?" (1984) in *Dits et écrits. 1954–1988*, vol. IV: 1980–1988. Paris: Gallimard, 1994: 568.
[5] Durand, Guy. *La Bioéthique. Nature, principes, enjeux.* Paris: Cerf, 1989: 19.
[6] See Toulouse, Gérard. *Regards sur l'éthique des sciences.* Paris: Hachette, 1998.
[7] http://www.nspe.org/Ethics/CodeofEthics/index.html (January 2010).
[8] Roqueplo, Philippe. *Entre savoir et décision, l'expertise scientifique.* Paris: INRA, 1997: 67.

CHAPTER 2

2 A brief history of space ethics

At the end of the 1950s, a French author wrote the following about innovation: "The passion that lives within the inventor has no relation, of any kind whatsoever, with its consequences. It is his personal reason for living, his own joy and his own suffering. His triumph over the provocative enigmas of nature is essentially personal. Whether this discovery is useful or perilous, fertile or destructive, it is of as little concern to him as the first rain. However, no-one is able to predetermine any of that. The consequences of a technical conquest of mankind are never predictable".[9] Should we, due to this, rule out any ethical questioning of the engineer's field of science, through fear of scaring away those men and women who practice science and stifling their passion? Is it enough to hide behind the screen of a so-called neutrality and to leave it to the experts and academics of the ethics committees to decide on the eventual use of discoveries, inventions and innovations? It is not that long ago that the answer to these two questions would have been affirmative, under the pretext that ethics could only slow down the progress of science and that it would therefore be advisable not to be hindered by it. Astronautics was no exception.

2.1 A slow emergence

However, the idea of an ethics of space emerged with the space venture itself. Many were doubtlessly unaware of this, including those within the astronautical community, but it was evoked by John F. Kennedy himself during the speech he gave at Rice University in Houston on 12 September 1962. Having summarised the then recent commitment of the American nation in the space race to land on the Moon, he explained to his citizens: "We set sail on this new sea because there is new knowledge to be gained, and new rights to be won, and they must be won and used for the progress of all people. For space science, like nuclear science and all technology, has no conscience of its own. Whether it will become a force for good or ill depends on man, and only if the United States occupies a position of preeminence can we help decide whether this new ocean will be a sea of peace or a new terrifying theater of war".[10] Is it not a concrete definition of space ethics?

Also in 1962, astronomer Bernard Lovell, who had been knighted by the Queen of England the year before, published a book entitled *The Exploration of Outer*

Space.[11] Having presented the investigation techniques employed by the astronomers of his time, the structure of the solar system and of the universe, its origin and evolution, he devoted a final chapter to "Some reflections on ethics and the cosmos"; what exactly did he mean by this? The English scholar begins by reminding us that the main elements required for life as we know it on Earth, in this case carbon and hydrogen, are found in abundance in the universe. How, then, could we avoid asking ourselves the age-old question regarding the possible plurality of inhabited worlds and the existence of possible extraterrestrial life forms? Moreover, as an astronomer, Lovell knew all about the West Ford Needles project. In 1961 (then in 1963), the US Air Force placed several million copper needles into orbit around the Earth at an altitude of approximately three thousand kilometres. The aim was to create a ring of space dipole antennae to act as passive reflectors for military communications. The first attempt failed and the second attempt was only a partial success. These needles then formed clusters. At the end of the 1990s, 65 of these clusters could still be observed from the ground. Lovell questioned whether it was acceptable to contaminate outer space in this way, to biologically contaminate planets other than our own, as the probes we send into space could potentially do. However, he concludes his last chapter and his book on an optimistic note: the race to the Moon and, more generally, the space race on which the Americans and Soviets were embarking provided excellent opportunities to direct the budgets of these two superpowers towards peaceful activities . . . .

Witnessing the birth and extremely rapid development of astronautical technologies, Pons in 1960 and Lovell in 1962, raised the question of ethics, deontology even, in their own respective ways. They realised that these technologies would offer, and were already offering, humans new possibilities and that it would be advisable to consider what place and purpose to allocate them. However, at that time, a more widely shared awareness was not yet promoted.

Twenty years later, in August 1982, at the second UNISPACE conference in Vienna, held to expand on international thinking in terms of space politics and law, several delegations expressed their concerns and their expectations. The Dutch delegate asked how it would be possible to put space at the service of developing countries, allowing them to acquire greater technological independence. He also raised concerns about the large share of space budgets still allocated to military activities. Despite the commitments undertaken by nations in the first Outer Space Treaty drawn up in 1967 (see below), the threat of the militarisation of space still seemed at this time to be a cause for concern. The Holy See representative, Msgr Peressin, also raised concerns about a possible transfer into space of terrestrial military confrontations: "[modern-day humanity] would need a very strong sense of solidarity and a firm desire to implement it at national and international levels. Unfortunately, these magnificent perspectives are overshadowed by the transfer into space of our military confrontations". The Greek delegate, Constantine Prevedourakis, defended similar ideas, focusing simultaneously on the mythology of his native culture and on the spirit of peace and cooperation that had been sweeping through Vienna since the end of the 1960s. Finally, Professor Soedjatmoko, rector at the University of the United Nations, invited to take a serious look at the crisis facing humanity: that of having reached the limits of its own planet. Would space not offer mankind a chance of survival, an opportunity to conquer new territories? However, to seize this opportunity would still require sufficient maturity to accept the need to find the best solution for everyone and to refuse to only support specific interests. He concluded: "Our generation is able to launch spaceships of the mind that may serve as pathfinders for epochs to come".

It was apparently the same line of thinking that inspired a UNESCO initiative the following year. The international organisation, headquartered in Paris, asked V. S. Vereschtin, Vice President of the "Intercosmos" Council at the Soviet Union Academy of Sciences, to prepare a round table on the subject of international cooperation in space. "Preserving space as a haven of peace and cooperation between the world's nations, and not allowing humanity to get accustomed to the idea that militarization of space is supposedly inevitable, wrote Vereschtin, is one of the chief objectives of law and ethics at the present time." The round table was held on 16 December 1983, as a teleconference bringing together six representatives from four continents. Peter Jankowitsch, then president of the Committee on the Peaceful Uses of Outer Space, stated with satisfaction that space provides humanity with a new vision of itself and, as such, leads to the creation of a new ethic. Despite this, he voiced some concern: "Whilst the first decades of the exploitation and use of space were characterised by a fortunate evolution of principles [of non-acquisition of rights of sovereignty in space and of celestial bodies, exclusively peaceful use in the interests of mankind] and whilst in particular the cooperation between the main space powers, the Soviet Union and the United States, have made great headway – the joint flight baptised *Apollo-Soyouz* constituting a spectacular step in this process –, the most recent years in space history are somewhat lacking in examples of international cooperation. On the contrary, several of the main principles pertaining to the use of space – and in particular the principle of an exclusively peaceful use of space activities – appear increasingly threatened". For his part, the American Isaac Rasool showed himself to be more sensitive to deontological questions such as the contamination of outer space and the status of data and information gathered. In agreement with Vladimir Kopal, Professor Bortzmeyer made a direct link between ethics and international law, in order to establish what he qualified as "organised and systematic development", whilst highlighting the necessity of imposing a general principle of governance alongside the declaration of outer space as common heritage of mankind. He also was anxious to turn technologies whose primary purpose often remained military to the benefit of peace.

Despite its relevant perspective and rich content, this meeting had no immediate concrete follow-up within UNESCO. It would take the UN organisation nearly 20 years to revive its interest in space. However, the initiative of December 1983 may have inspired the meeting held in Casablanca in March 1984, under the aegis of the Royal Moroccan Academy and the following title: *De la déontologie de la conquête spatiale* (The deontology of the conquest of space). Going through the presentations of this event, it is impossible not to be struck by the lucidity of the participants. Whilst most of them appeared still concerned by the threat of an increasing militarisation of space (President Reagan launched his Strategic Defense Initiative, often referred to as Star Wars, in March 1983), all the same, they did not forget the other challenges arising from space activities. In addition to the challenges arising due to technological expertise (the congestion of orbits and the proliferation of debris, the pollution caused by launchers, etc.) come legal and diplomatic issues (which sovereignty? how is data shared?). One of the participants at this meeting asked whether the "Earth's damned" would be joined tomorrow by "Space's damned" following difficulties, even the impossibility of Third-World countries accessing data and space technologies. How would it be possible not to feel excluded or even dispossessed, faced with the arrogance of the space powers and prevailing inequality? The participants of the session in Casablanca recognised that one of the key issues was not so much to specify or to revive the spirit of Vienna, the letter of international agreements having already been drafted and signed, but to put them into practice, in other words to define an explicit deontology and to ensure its application. How could this exploit be a success with no coercion possible and relying solely on good will? There is also the question of the consequences of the use of space resources on cultures, on their spreading or on the contrary on their isolation or even their disappearance. More generally, is it possible to consider or to claim that space has already led to and may still lead to true innovations in the sociocultural domain?

We can conclude that at the start of the 1980s a wave of concern spread through the international space community with regards to an ethical and deontological questioning, partly with regards to the outlook revealed by legal work: the UNISPACE II conference in Vienna, the UNESCO teleconference in Paris, the meeting organised by the Royal Moroccan Academy in Casablanca may be considered as the successive stages of a rising awareness which by no means remained at a superficial level but rather resulted in the definition of the main focus points and the essential view points for future space ethics. Unfortunately, save the discourses and acts published by the Moroccan institution, nothing arose from or was retained of these three initiatives. Although the Japanese space agency NASDA conducted a study into the cultural consequences of space activities on Japanese society in the late 1990s, it was another fifteen years before ethics was granted any significance within the space community.

"At the initiative of the Director-General of UNESCO, Mr. Federico Mayor, and acting on a proposal by the Director-General of the European Space Agency (ESA), Mr. Antonio Rodotá, a new working group was set up to consider the ethics of outer space in December 1998 on the basis of a partnership between UNESCO and ESA". It is in these terms that the report published by UNESCO in July 2000, under the title *The Ethics of Space Policy*, explained the origin of the renewed interest of the space sector in ethics: a proposal from the European Space Agency and a UNESCO initiative. Coordinated by Professor Alain Pompidou, the working group[12] attempted "to identify the difficulties and fears, opportunities and promises associated with the conquest of space, while providing the necessary explanations in the clearest and most comprehensive manner possible, taking account of the needs of the populations in their specific socio-cultural context". From this report, I will take a triple definition of space and one notable omission. Space is presented as a dimension, in other words a place, an environment; as a tool, due in particular to the communication and observation satellites; and finally as a perception as held by the general public and portrayed by the media.

The notable omission is the coverage of military activities. The strictly civil nature of ESA projects and programmes was incidentally the main reason given by its managers to explain its last-minute withdrawal from a conference held around the same time, in March 1999, at the Darmstadt University of Technology in Germany. The organisers of this event had set aside a significant amount of time to discuss the question of the military use of space technologies, space dominance and the weaponization of outer space.[13]

Following this report, UNESCO set up within its World Commission on the Ethics of Scientific Knowledge and Technology (COMEST[14]) a sub-commission devoted to outer space. This sub-commission led several actions, mostly with the support of ESA: reports, conferences, etc. Since 2005, it seems to have become inactive, without doubt due to redundancy or competition with another United Nations organisation, the Committee on the Peaceful Uses of Outer Space (COPUOS, see below). In June 2001, following the publication of the *Ethics of Space Policy* report by COMEST, COPUOS in fact devoted one of the sessions of its annual conference in Vienna to the topic of space ethics. This was no doubt a way to mark its territory, a reminder that it had been the first to explore the field of space ethics, due to its legal expertise.

At the same time, the French Space Agency CNES, adopted a similar process. At the start of 1999, Gérard Brachet, the Director General of the French space agency, entrusted a group of engineers with the task of exploring the field of space ethics. Less than three years later, in October 2001, a book was published to divulge the fruits of this exploration, entitled *Icarus' Second Chance*. Six months earlier, the management at CNES had created the post of Ethics Adviser. The role of the appointed adviser would be to continue the work started by this group of CNES staff.

In January 1999, notified of the study undertaken by ESA and UNESCO, the famous French scientific journalist Albert Ducrocq used his column published in the *Air et Cosmos* review to voice his opinion concerning these space ethic initiatives. He supported the idea according to which space debris presents practically no danger to Earth. In reference to discourses about Mars samples, he added that they only provoke fear among the public and detract from the scientific interest of such a mission. He concluded: to adopt an ethical stance, to draw up a sort of space charter "risks giving credit to the argument according to which, not content with wasting resources, astronautics is harmful enough to put the planet in danger". In other words, the innocent intention of ESA could "be highly detrimental to the case for space".[15] Ducrocq's position is worth focusing on. He believes there are enough arguments to defend the idea according to which space would largely avoid the ethical questions with which other technological domains are today confronted. In other words, there would be a space specificity in this domain.

No doubt an echo of a few members of the astronautical community, the reaction of the renowned French journalist merits attention and respect. However, due to the initiatives of ESA and CNES, his reaction did not prevent ethics from taking firm root in the field of space. Organisations such as COPUOS devote seminars to the subject; conferences organised by the International Astronautical Federation (IAF), by the International Academy of Astronautics (IAA) and the Committee on Space Research (COSPAR) welcome discussion on the theme of space and ethics. the International Space University (ISU), located in Strasbourg (France) honours its intercultural and interdisciplinary dimension by running its summer school and its master's in ethical issues. Studies assigned to the European Science Foundation (ESF) by the European Space Policy Institute (ESPI) concerning the future of space activities also touch on this new field. However, CNES is the only space agency to have appointed individuals or a team responsible for the ethical dimension of their activities and even, as I have already observed, the interest of ESA would appear to need a new impulse. In terms of space, ethics still remains a frontier to be crossed.

2.2 Lessons to be learned

Ten years after the publication of Professor Pompidou's report and the start of the work by CNES in terms of ethics, what are the lessons that can be learned from this period?

The first takes the form of an observation: this approach has sparked and still sparks much interest among those to whom it is presented. The initial reaction is often one of surprise. "Space ethics? Never heard of such a thing!", "I have absolutely no idea what that could be about . . . ". Many of our contemporaries would sum space up as a series of spectacular events, with high media coverage, which apparently have nothing to do with what the word ethics normally designates: medical research, financial practices, academic councils, etc. However, once our contemporaries have been shown the cultural roots of space conquest and the social issues of these activities highlighted, most of them recognise and appreciate the approach adopted in terms of ethics. They are attune to this return to, or simply this focus on, the human factor within space activities and achievements, a view that is not just limited to the presence of astronauts, to manned flights.

A second lesson concerns the implementation of an ethical approach within a professional environment. At the start of the 2000s, those passionate about space ethics sought to multiply the opportunities to talk about it and promote it. They wrote articles and books, attended colloquiums, organised sessions during international conferences, and gave interviews in various professional reviews, which was without doubt the best approach to adopt. However, a dwindling interest, even total lack of interest, soon occurred, posing a threat. What might be the reason, what could be the solution? I actually like telling my colleagues at CNES: "If you are a competent engineer or a committed scientist, you will in fact be practising ethics all the time, without realising it . . . ". It is not therefore a question of introducing concern for ethics to their professional activities because it is lacking, but moreover a question of demonstrating the already effective presence of this concern, to further highlight it, to share it and to expand on it.

Therefore, those responsible for or with an interest in space ethics must see beyond long *ex cathedra* dissertations and overly theoretical or academic thoughts and promote its integration into the practicalities. Priority must be given to what those who act have to say, before questioning them and asking questions about the meaning, means and consequences of their practices. Now, of the conferences to which I would be sent by CNES on my own to talk about space ethics, I prefer those at which a colleague first takes the floor to speak about future planetary exploration missions, of the management of space debris, of the possibility of exploiting lunar resources or of the UFO study group, before sharing my questions and opinions with the auditorium, from my own ethical perspective and experience. I am convinced that this is the right way and the right point at which to bring up the topic of ethics.

A third lesson can be learned from the response to a question often posed to CNES: why, in 2001, did the management team not appoint an ethics committee, as several French scientific research and technological development organisations had already done? The reasons are as follows. Firstly, because the subjects addressed by space ethics do not strictly speaking constitute "cases" or "files" to which it would be possible to assign a group of experts or academics. It is more a question of topics to be studied, to be researched over the long term. Secondly, if CNES were confronted with a real ethical issue (as genetic engineering researchers, doctors, etc. are), it could submit it to the ethics committee of one of these other organisations (e.g. CNRS) with which it often collaborates. Is it not precisely one of the characteristics of the space venture to implement and maintain permanent and effective links with a number of scientific and technological fields? For all that, it should not be concluded that space ethics has no specific or unique character. On the contrary, it provides us with what is simply and yet appropriately referred to as the "point of view of Sirius". It is as if space involved reaching both a physical and an intellectual height, as if it offered the chance to introduce a different perspective in the intuitive understanding that we have of the reality, issues and challenges it proposes or imposes. I will now give a few illustrations of this ownership of space.

2.3 Space law, a heritage in progress

In the brief history of space ethics given above, I purposefully omitted a whole branch: space law. Theories and opinions diverge concerning the place to be granted to the legal field in relation to ethics. Nonetheless and with regards to the modern history of space, I do not hesitate to associate law and ethics very closely: the premature genesis of space law profoundly marked the slow development of space ethics, by creating several key fundamental principles and by having them recognised by the international community, by establishing competent structures integrated on an international scale. In other words, space law appeared as the oldest expression of space ethics, in a specific geographical, historical and cultural location.

2.3.1 Genesis of space law

Space law is essentially based on international treaties drawn up during the 1960s and 1970s, at a time when space activities were those of a monopoly of a few space powers, first and foremost the United States and the Soviet Union. However, the founders of astronautics had already envisaged the establishment of a legal system specific to the field of space. Following the works of Konstantin Tsiolkovsky, E. Laude focused people's attention, as early as 1910, on what he called cosmic law as a different and independent legal branch of air law. Later, at the time when the V2s, retrieved in Germany at the end of the Second World War, were used to conduct the first scientific experiments in outer space, the research community also started to ask questions about the necessity and content of a space law. When, on 4 October 1957, during the International Geophysical Year, the Soviets launched Sputnik 1, no State protested against being flown over by the Soviet satellite, without prior authorisation. The Soviets raised the idea of a tacit agreement of States, in order to successfully conduct the meteorological, magnetic and electrical exploration of the upper atmosphere to which that year was devoted. Others regarded this event as a spontaneous acceptance of the principle of the freedom of the use of space, a principle that would later be evoked when drawing up international treaties. Meanwhile, the US Department of State requested the regulation of space activities, emphasising the exceptional and transient nature of the "argument" of the International Geophysical Year.

In fact, the principle of a space treaty that would be placed under the aegis of the United Nations was evoked for the first time in 1956. The year after and following the launch of Sputnik, several studies were conducted. The International Council of Scientific Union (ICSU) looked at the scientific aspects of the space conquest, whilst the International Astronautic Federation (IAF) established the Permanent Committee of Space Law in 1958 (later known as the International Institute of Space Law, IISL, in 1960). During this time, the United Nations adopted, on 14 November 1957, resolution no. 1148 recommending the pursuit, within the context of disarmament, of "the joint study of an inspection system which would only allow objects to be sent through outer space exclusively for peaceful and scientific purposes".

In 1958, the UN set up the Committee on the Peaceful Uses of Outer Space (COPUOS). Until 1962, the first agreements between the United States and the Soviet Union aimed to ensure that States did not disrupt the space activities of other States. It was not until 1965 that France became the third space power. COPUOS then created two subcommittees, one legal and the other scientific and technical. The legal subcommittee managed to fairly quickly compile a text aimed at regulating the legal problems associated with the peaceful use of space and the principle of non-appropriation. The scientific and technical committee was notably put in charge of the two questions dealt with in this book: the management of space debris and planetary pollution. Alongside COPUOS, other specialist UN institutions indirectly assisted and continue to assist with the elaboration of a space law: UNESCO, the International Telecommunications Union, the World Meteorological Organization, the International Civil Aviation Organization. Intergovernmental organizations specialising in space activities were also involved in this task; these included Intelsat (a consortium set up in 1964 which became a private company in 2001), Intercosmos (of Soviet origin) and ESA.

The principles of space law are set out in the *Treaty* on *Principles Governing the Activities of States* in the *Exploration and Use of Outer Space, including the Moon and Other Celestial Bodies*, or Outer Space Treaty, dated 19 December 1966. It was signed on 27 January 1967, first ratified on 10 October 1973, and has today been ratified by 97 States and signed by 27 others. These principles were then completed and developed by other international texts:

- the *Agreement on the Rescue of Astronauts, the Return of Astronauts and the Return of Objects Launched into Outer Space*, concluded on 22 April 1968;
- the *Convention on International Liability for Damage Caused by Space Objects*, concluded on 29 March 1972;
- the *Convention on Registration of Objects Launched into Outer Space*, concluded on 14 January 1975;
- the *Agreement Governing the Activities of States on the Moon and other Celestial Bodies*, concluded on 18 December 1979.

Several resolutions adopted by the United Nations General Assembly also need to be added to this list of agreements and conventions. These include the resolution of 3 December 1986, pertaining to remote sensing principles (no. 41/65) and the resolution of 14 December 1992, pertaining to the use of nuclear power sources in outer space (no. 47/68). To present day, these resolutions have been implemented by the member States.[16]

2.3.2 Responsible States

The responsibility of States in terms of space activities constitutes a significant aspect of space law. This question is principally dealt with in the 1972 Liability Convention although it was already detailed in Article VI of the 1967 Outer Space Treaty: "States Parties to the Treaty shall bear international responsibility for national activities in outer space, whether such activities are carried on by governmental agencies or by non-governmental entities ... " In other words (and this is what makes space law so unique), all "national" space activities are assimilated to activities of the States, whether carried out by their citizens or by foreigners in their country, by their nationals from the territories of other States or even from international zones. By virtue of Articles VI and VII of the same Treaty, the responsibility of States Parties shall be understood according to two meanings: that of the control of the actual activities (responsibility) and that of financial compensation in the event of damage caused by these activities (liability).

The 1972 Liability Convention specifies the scope of application of this specific rule for the compensation of damage caused by a space object on the ground, in air space and in outer space, based on the notion of "launching State". The first Article of this Convention defines this notion as follows: "The term 'launching State' means: a State which launches or procures the launching of a space object; a State from whose territory or facility a space object is launched". In other words, this notion essential to space law is defined based on four alternative criteria: the State which launches, the State which procures the launching, the State from whose facility a space object is launched and the State from whose territory a space object is launched. In this way, for a single object launched into outer space, several countries may be recognised as "launching States" and therefore be jointly responsible for damage which may be caused by this object: "a launching State shall be absolutely liable to pay compensation for damage caused by its space object on the surface of the Earth or to aircraft in flight". Article III of the convention specifies that "in the event of damage being caused elsewhere than on the surface of the Earth to a space object of one launching State or to persons or property on board such a space object by a space object of another launching State, the latter shall be liable only if the damage is due to its fault or the fault of persons for whom it is responsible". No time limit is specified in terms of this liability, which in itself poses a new legal problem as objects launched into orbit may remain there for several centuries. During this period, the space objects may change ownership and, consequently, launching State (or a new launching State may become involved). They may also be moved repeatedly without the launching State(s) being informed despite the fact that the latter remain legally liable.

The 1972 Convention goes on to specify several pertinent points, in the event of accident:

- determination of the launching State or States in the event of a joint launching. In this latter case, the launching States may conclude agreements regarding the apportioning among themselves of the financial obligation in respect of which they are jointly and severally liable;

- damage claim procedure and possibilities of enlisting a Claims Commission;
- joint liability: the victim may turn to the most solvent of the launching States to demand the full amount of compensation, the latter being able then to demand funds from the other launch States, if the distribution of the financial liability has not been established between them in an agreement;
- composition and running of the Claims Commission, etc.

Commonly used on Earth, the notion of fault is not easy to use when applied to space activities as it explicitly refers to a regulation which, in most cases, does not exist or is difficult to apply as it exists in several forms and under several designations. This is, in particular, the case of space debris in circumterrestrial space which has led to the drawing up by space agencies of several types of recommendation: the *Mitigation Guidelines* of the Inter Agency Space Debris Coordination Committee (IADC), the NASA standard and the European code of conduct. The way in which these texts define "good behaviour" in space leads to several different and sometimes even opposing points of view.[17]

The notion of fault also raises the question of evidence. How is it possible to gather evidence when the activities in question take place hundreds and thousands of kilometres from the Earth's surface, in zones that are difficult to access? Let us return to the example of space debris. The most readily available information about objects moving around in space is obtained via American space surveillance systems (see below). NASA regularly publishes and circulates a catalogue of objects moving around our planet. This document is of course declassified, of limited precision and refers only to objects which the United States considers to be pertinent. How credible is this information when attempting to investigate an incident to determine fault and liability? One-off or partial observations conducted by other nations have allowed the errors, mix-ups and even omissions of the American services to be highlighted whether these were unintentional (the identification of small and far-off objects is difficult) or deliberate. How can we therefore hope to provide tangible, solid evidence in the event of a dispute?

It goes without saying that, in space, fault and proof are very difficult to establish. Even so, the application of the singular notion of launching State must not be only *a posteriori*, i.e. after the incident or accident, but *a priori*. It can then become a tool, a political lever that allows us to ask questions that should be inevitable: what rules need to be applied? To whom does the satellite really belong: the customer, the insurers? Who makes the decisions?

Finally, the question of the implementation, monitoring and control of this law comes. We cannot simply define regulations concerning the use of space. We must be able to check the compliance with these regulations by the different players and operators. This requires the implementation of independent means of observation. Then, if a breach is duly observed, e.g. a violation of the principle of the use of space for peaceful purposes, a sanction would need to be applied. However, is it possible to establish such a system of space policing at an international level? More realistically, it is now recommended, notably by COPUOS, that we ensure the national application of rules defined at international level. So, each State is requested to implement a national legislative system in compliance with international space regulations. There is a risk that we could find ourselves in a situation identical to that of the law of the sea where the practice of flags of convenience exists. A State could in effect choose not to impose the application of international measures on its inhabitants and the companies on its soil and to thus favour the implantation on its territory of operators more concerned with the reduction of their costs than common good. For the time being, the main space States have signed and do implement the UN treaties. However, the risk of flags of convenience turning up in space cannot be ruled out.

In April 1995, the Sea Launch company was set up. It uses the three-stage Russian launch vehicle Zenit-3SL from the floating Ocean Odyssey platform, built based on the Norwegian technology of off-shore oil drilling platforms. The platform is installed as close to the Equator as possible, near Christmas Island, in order to minimise launch power. The Seattle-based Boeing Commercial Space Company owns 40% of the company, the Moscow-based company RSC-Energia owns 25%, the Anglo–Norwegian company Kvaerner Maritime a.s. owns 20% and the Ukrainian company SDO-Yuzhnoye/PO-Yuzhmash owns the remaining 15%. With its headquarters initially in the Cayman Islands, Sea Launch recently opened its offices in the USA, in the State of Delaware, and has offices in Seattle and in Oslo, Norway. The port of registry for the command and assembly ship, and the platform is Long Beach in California and both ships are registered in Liberia. They belong to two different companies, subsidiaries of the Sea Launch Company. It is hardly difficult to imagine the legal quagmire that the smallest claim or complaint would provoke, in the event of an incident or accident, in the context of such a complex set up.

2.3.3 The law retranscribed

Some States, including the major space powers, have adopted legislation aimed at governing their national space activities and thus at implementing their international obligations. These States include the United States, Russia and Australia. European countries are progressively following in their footsteps. Following the

United Kingdom in 1988, Belgium adopted a space law in 2005 and other laws are also being drafted in Germany, Italy, Luxembourg and Switzerland. France is also up there, as a space law was voted in June 2008. Following the example of other national laws, the primary vocation of the French law consists in implementing rules to authorise and control national space activities, particularly those undertaken by private companies, in compliance with obligations arising from international treaties. These rules are very much a necessity as the French State is heavily involved in space activities. In effect, France is home to the CSG (Centre Spatial Guyanais, the Europe's Space Port located in French Guyana), making it a launching State for all Ariane launchers and now also for Vega and Soyouz launchers. It is likely to be liable for all damage caused by a space object launched from this base, whether by the launch vehicle or its debris or by the launched satellite. Moreover, the major European space operators, whether Arianespace or Eutelsat, are companies governed by French law. In this case, the law sets out an important role for CNES as far as the implementation of this law is concerned, notably entrusting it with testing the technical compliance of national space systems.

Up to now, France has been able to manage national space activities and ensuing risks predominantly via conventions, i.e. by constraints, in particular technical, imposed in the agreements concluded with the main players in the space sector, principally ESA and Arianespace. These constraints include CNES doctrine of back-up, a collection of technical rules drawn up by the French space agency, which aims to manage the technical risks created by risky activities taking place at the CSG. French space law reinforces the role of CNES in the matter, expressly entrusting it with the general task of protecting persons, property and the environment, on ground and in flight, concerning the CSG.

Now its history has been recounted and its legal context set out, we are at last ready to tackle the subject of space ethics itself. It is important to remember that space ethics is fundamentally and always human. It is part of the oldest dreams and myths, those same dreams which drove the human species from the lands of Africa to the territories of the sky.

[9] Quoted in to Arnould, Jacques. *La seconde chance d'Icare. Pour une éthique de l'Espace*. Paris: Cerf, 2001: 84–85.

[10] *Public Papers of the Presidents of the United States*. 1962. 1: 669–670.

[11] Lovell, Bernard. *The Exploration of Outer Space*. New York: Harper & Row, 1962.

[12] Alongside Professor Pompidou, this group includes five members whose names and job titles are as follows: Jean Audouze, astrophysicist, Research Director at the CNRS, Director of the Palais de la découverte, Ezio Bussoletti, Professor at the Naval University of Naples, Director of the Experimental Physics Institute; Carl Friedrich Gethmann, Professor at the Institute of Philosophy of Essen University, Director of the European Academy for Study of the Consequences of Scientific and Technological Progress; André Lebeau, Member of the National Air and Space Academy, Professor at the Conservatoire National des Arts et Métiers (CNAM), former Deputy Director-General of the European Space Agency, former President of the French National Space Research Centre (CNES) and Sir Geoffrey Pattie, Director of Communications of The General Electric Company.
[13] See *Space Use and Ethics.* Ed. Wolfgang Bender, Regina Hagen, Martin Kalinowski and Jürgen Scheffran. Volume I: Papers. Münster: Agenda Verlag, 2001.
[14] Acronym taken from the French name "Commission mondiale d'éthique des connaissances scientifiques et des technologies".
[15] Ducrocq, Albert. "Ethique spatiale. Une bonne intention qui pourrait fort mal servir la cause de l'espace". *Air et Cosmos.* 1686 (1999): 39.
[16] See *Text and status of treaties and principles governing the activities of States in the exploration and use of outer space, adopted by the United Nations General Assembly A commemorative edition, published on the occasion of the Third United Nations Conference on the Exploration and Peaceful Uses of Outer Space (UNISPACE III).* UN: Vienna, 1999.
[17] Alby, Fernand *et al.* "A European standard for space debris". *Air and Space Academy Colloquium "Europe and Space Debris"*: Toulouse, 27 and 28 November 2002.

CHAPTER 3

3 Icarus

When it came to me to chair the first meetings of the group appointed by Gérard Brachet to explore field of space ethics within CNES, I had to overcome the surprise and concerns of my colleagues at having to address a philosophical subject for which their engineering training and tasks had not prepared them. To do that, I turned to one of the oldest methods used by our ancestors to ask questions about the meaning of life and the task of being human: mythological narrative. The narrative of Icarus sprung to mind immediately.

3.1 The myth of Icarus revisited

"I warn you, Icarus, to fly in a middle course, lest, if you go too low, the water may weight your wings; if you go too high, the fire may burn them. Fly between the two. And I bid you not to shape your course by Boötes, the Herdsman, or Helice, the Great Bear, or the drawn sword of Orion, but fly where I shall lead".[18] Not too low, not too high, not too far north, not too far south. The words of advice spoken by Daedalus to his son Icarus are about taking care and about moderation. Were they not the first two humans to use wings to fly? Daedalus (his name meant "the craftsman") knew from experience that the genius of man was not without setbacks and could lead to the most crazed and immoral acts. Had he not in the past assassinated his own nephew and disciple who, rivalling his master's skill, had invented the sword and the compass? The young man, thrown from the top of Minerva's sacred citadel, was transformed into a partridge. Daedalus showed deep concern when he saw his son learn to use the first wings for flying: "He kissed his son, which he was destined never again to do, and rising on his wings, he flew on ahead, fearing for his companion, just like a bird which has led forth her fledglings from the high nest into the unsubstantial air".

Modern interpretations of the story of Icarus often viewed it as the myth inspiring aeronautical and astronautical adventure, the conquering of the sky by man now become equal to birds, or even gods. The myth warns us however to be careful. A fall as spectacular and fatal as Icarus' awaits all those who let themselves be carried off by *hubris*, the arrogance of the creative genius. Ovid continues his narrative: "The scorching rays of the nearer sun softened the fragrant wax which held his wings. The wax melted; his arms were bare as he beat them up and down, but, lacking wings, they took no hold on the air. His lips, calling to the last upon his father's name, were drowned in the dark blue sea". Interpreted in this way, the myth of Icarus is magnificent: it founds one of the most modern human momentums whilst imposing physical, psychological and even ethical boundaries. But, is that really all that this myth has to tell us? I am not so sure.

To start with, I propose to re-establish two truths about the mythological narrative itself. The first truth concerns Icarus, son of Daedalus. Icarus is often portrayed as being a handsome and athletic young man, rather like a Greek statue, as if it were necessary to have this appearance to conquer the sky and fall due to excessive arrogance. However, Ovid's narrative says nothing of the sort. It evokes the young Icarus, then the child to whom he does not directly attribute a feeling of arrogance, but rather more the experience of pleasure: "And now Juno's sacred Samos had been passed on the left, and Delos and Paros; Lebinthos was on the right and Calymne, rich in honey, when the boy began to rejoice in his bold flight and, deserting his leader, led by a desire for the open sky, directed his course to a greater height". I previously spoke of moderation, about which Daedalus advised his son. However, is a child capable of understanding such advice? Disobedience is not necessarily synonymous with pride and caution is not always inherent in youth. Those who read in the myth of Icarus a warning against human pride are probably not wrong. However, perhaps it could also be interpreted as a simple call to good sense and caution, to preclude the dangers that go hand in hand with being care-free and young. It may be a less moral discourse but is certainly more realistic. That said, the advice Daedalus gave to his son conveniently complements the often repeated words of Konstantin Tsiolkovsky: "Earth is the cradle of Humanity; but mankind cannot stay in the cradle forever"[19] and yet: "Mankind will not remain on Earth forever [. . .] but will at first timidly penetrate beyond the confines of the atmosphere, and later will conquer for itself all the space near the sun". Daedalus and Tsiolkovsky both speak of "Caution", as the first steps of mankind in space are those of both a child and a conqueror. The second lesson to be learned from the myth of Icarus corroborates the first remark.

What is in fact the reason for this mythical flight? The conquest of the skies? The acquisition of an extreme power? Not at all. If Daedalus designed and built the wings by observing and copying those of birds (as Leonardo de Vinci would do later, at least theoretically), it was to escape the prison in which he was incarcerated by King Minos. Escape: there lies the real reason for the aerial adventure of Daedalus and Icarus. The two ancestors of Wright and Ader, Gagarin and Leonov, Armstrong and Aldrin were not proud *conquistadors* but humble yet ingenious and brave runaways. This other interpretation of the myth of Icarus, whilst it may be less flamboyant, is no less possible. Is escape not the reason given by supporters of the "space option"? Individually or collectively (one simply needs to browse the website and publications of the American *National Space Society*[20]), they actively support a reinforcement, even an increase in space activity globally, in particular through the impetus and with the support of state funding. They base their argument on the observation of the deteriorating conditions of life on Earth and the exhaustion of the Earth's raw material, energy and vital space resources, as described in the numerous reports following that of the Club of Rome in 1972. They defend not only the need to implement the means to explore extraterrestrial resources, but also the need to set about colonising the solar system to escape the harmful consequences of overpopulation and ensure the survival of mankind. The only way to ensure mankind's survival is to escape into space. I am convinced that these space option activists are the most direct descendents of Daedalus and Icarus.

Let us continue our reading of Ovid's narrative and, without forgetting the two aforementioned interpretations, give it a resolutely modern twist. To do this, let us start from the beginning, in other words with the very reason why Daedalus built a labyrinth at the request of King Minos. Pasiphae, the wife of King Minos, fell in love with a magnificent bull appearing from the waves upon the order of Poseidon. She asked Daedalus to build her a wooden cow in which she hid: of the magnificent beast deceived by the subterfuge (still used by inseminators today), she gave birth to a son, the Minotaur. Ovid described the Minotaur as a hybrid beast: a man with a bull's head. Wronged husband but apparently not one to bear grudges, Minos once again turned to Daedalus for help asking him to imprison this dreaded beast "in a labyrinthine enclosure with blind passages", in a labyrinth so carefully constructed that the architect himself had difficulty finding his way out. Theseus later used Ariadne's thread to escape from it. Later held prisoner on the island of Crete by Minos, Daedalus decided to escape: "Though he may block escape by land and water, he said, yet the sky is open and by that way I will go. Though Minos rules over all, he does not rule the air". He then began to build wings, for him and his son.

With implications beyond the tragic death of Icarus, the myth of Daedalus is strangely echoed in modern technoscience. Let us start with the hybrid Minotaur born of an unnatural relationship: although not on a par with the creature of Doctor Frankenstein, can the Minotaur not be equated with a type of prototype of organism now obtained using biological techniques using either genetic engineering, selection or crossbreeding techniques? Unlike Minos, the husband cheated on by his wife, we feel more fear than shame, to the extent that we also try to keep these organisms contained and prevent diffusion and proliferation using mechanical or biological labyrinths. However, as descendants of Daedalus, are we not in danger of being trapped in the prisons which we have ourselves built? There are vicious circles that are as effective as Daedalus' labyrinth. The mass use of antibiotics is one example. One of their inventors, René Dubos, even theoretically specified the rules and limits. Let us return now to the conclusion of the myth of Icarus and the wings that burned and the wax that melted due to his closeness to the sun. I am of course extrapolating, but why should we not interpret in this episode and in our modern language, a sort of lesson concerning our management of energy, in particular nuclear energy? It is clearly not a question here of taking sides in the debate between supporters and opponents of nuclear energy, but simply of noting how fears about it are expressed in similar terms, which raise questions concerning the outrageousness of humans and emphasise their lack of caution. This is the role that has always been given to myths.

Trapped between the Minotaur and the sun, between GMOs and nuclear energy(?), the flight of Daedalus and Icarus no doubt caused, fleetingly, the stupor and surprise of the fisherman, the shepherd or the labourer who witnessed the man's first flight. Yet, when Icarus crashed into the sea which now takes his name, only his father was concerned: the labourer continued undisturbed in following his trail, at least if we believe the famous paintings of Pieter Bruegel the Elder on this ancient theme, *Landscape with the Fall of Icarus*. While this might not be a satisfactory illustration of space venture, it is at least an honest one. Next to genetic engineering and nuclear technologies, it does not instil enough fear to spark any ethical commotion, even upon the Challenger shuttle disaster in February 1986 resulting in the death of its seven crew members, in the Atlantic ocean or when, seven years later, the Columbia shuttle burst into flame upon re-entry into the Earth's atmosphere. Following both disasters, the American authorities appointed an enquiry commission, but not an ethics committee.

3.2 From dream to reality

The myth of Icarus is only an illustration. Space constitutes one of the domains on which the human imagination feeds and expresses itself. Our fascination with the sky is such that the cultures of all countries and of all eras have not only placed their gods, their paradises and their origins there but also meaning, the destination of their desires and their dreams. Yet, access to space was beyond man's reach for a long time. Although science, aided by technology, explained very early on "how the heavens go", to use a favourite expression of Galileo, it left to the domain of religion and spirituality, on the one hand, and to the domain of imagination and fantasy, on the other hand, the task of demonstrating and teaching us "how to get there". These two domains are probably not totally independent but I will nevertheless not deal with the relationship of humans with the heavens, preferring to leave that to the field of religion and the sacred. Instead, I will just deal with what can be qualified as an imaginary journey into space.

The first written trace of such a journey dates back to the year 180AD and this was Lucian of Samosata's text entitled *A True Story*. Lucian informs his readers that he writes of "things I have neither seen nor experienced nor heard tell of from anybody else; things, what is more, that do not in fact exist and could not ever exist at all. So my readers must not believe a word I say" He then tells the story of a ship whipped up into the air by a violent storm, his journey through the skies for seven days and seven nights, his arrival on a mysterious cosmic island, his meeting with strange creatures, etc. In *Icaro-Menippus*, Lucian writes about another cosmic journey and how a man with two wings travels to the Moon and beyond. Since Lucian, many authors have explored this literary genre. At the beginning of the 14th century, Dante Alighieri proposed a tour of the seven spheres of the cosmos, guided by Beatrice. In 1638, Francis Godwin published his vision of a wildly enchanting lunar nature and of a people more successful than our own, in *The Man in the Moon*. Around the same time, Cyrano de Bergerac published *Les Etats et Empires de la Lune* (The states and empires of the Moon, 1657) followed by *Les Etats et Empires du Soleil* (The states and empires of the Sun, 1662). In 1765, Marie-Anne de Roumier published the seven volumes of *Voyages de Milord Céton dans les sept planèts* (Lord Seton's Voyage Among the Seven Planets): a true astronomical epic. In 1835, Edgar Poe sent *Hans Pfaal* to the Moon on board a balloon.

The 19th century marked the end of the systematic exploration of the terrestrial globe led by the West since the end of the 15th century. Space presented itself as a new field of expansion: space travel was no longer just a theme in stories, it became scientific. Although it had already been around for many years, the astronomical telescope became very popular. Combined with a good dose of imagination, the telescope enabled astronomers to view the Moon, planets and stars. They were even able to produce a map of the surface of Mars that was as accurate as maps of the Earth, or so they claimed. Camille Flammarion and his numerous publications were the talk of the end of the 19th century. In his book *La Pluralité des mondes habités* (The Plurality of Inhabited Worlds), the author claimed that "the Earth has no marked supremacy in the solar system to make it the only inhabited world, and, astronomically speaking, other worlds have just as much potential as ours to offer a home for life".

There was no shortage of imagination and authors in the 20th century: from Herbert G. Wells (*The War of the Worlds* in 1898, *The First Men in the Moon* in 1901) to Arthur C. Clarke (*2001, Space Odyssey* in 1968), the literary sphere of space fiction became very popular. Wells can be considered as the inventor of modern science fiction. He was responsible for the public knowledge of "scientific wonders" such as the time machine, the element transmuter and hyperspace. It is sometimes forgotten that Clarke was a member of the British Interplanetary Society: in the 1930s, he had designed with others members of this society a lunar mission, before coming up with the concept of geostationary satellites. The list could go on.... Until the mid 20th century, the sky, which, in the West, had gradually lost its religious character, then became the domain or the reason behind a powerful work of imagination.

4 October 1957 marked a turning point for space adventure. It was no longer the product of the fruits of the imagination or astronomical research alone. With the launch of the first Sputnik, space adventure entered the era of technological production, success and conquest. The cosmic journey, for which the West had been making imaginary preparations for such a long time, became achievable, though perhaps not immediately. Mankind could now pride itself on its title of citizen of the cosmos. Even so, space fiction continued to exist and in his book *Space and the American Imagination*, Howard McCurdy demonstrated how authors and promoters of the United States space programme relied on well-known cultural figures such as Walt Disney to arouse the interest and support of the American public and its imagination.[21] Not exclusive to America, this link between dream and reality can also be found in other countries and other cultures. Japanese interest in lunar missions could be explained by the popularity of Princess Kaguya in Japan. This descendent of the Moon people had been found in a bamboo plant and was thus named *Naotake no Kaguya Hime*, the radiant princess of the bamboo. Much to the annoyance of the emperor who had wanted to marry her, her father helped her return to her native Moon.

At the same time, space continued to feed the imagination and to fill the pages and scripts of science-fiction novels and films, a creative field in which it has never been difficult to find food for ethical thought. Despite being fictional, "encounters of the third kind" provoked questions about the respect accorded to the other, whatever their nature and however different they might be. Likewise, the societies populating the Earth for millennia to come or the planets in deep space are reflections of our ideas and our fears, in terms of government. Let us not forget themes such as that of the status of new conquered worlds or that of progress or the dangers provoked by sciences and future technologies, whether foreseen or imagined.

Why this lengthy digression about fiction and its fixation on space? Because myth and fiction go hand in hand. Let us return briefly to the myth of Icarus and to my suggested interpretation. Contrary to what a rationalist and simplistic interpretation would consider, this myth is not just a work of fiction, a beautiful story, intended to allay human conscience and reasoning; or it could not claim any part in the elaboration of human ethics. Psychological, sociological or ethnological disciplines exist to unearth, highlight and analyse, within the mythologies of our species and our cultures, the processes that give them a status and a role much more important than that of a "beautiful story". According to the words of Dominique Lecourt, we need to definitively accept "all narratives which challenge and put to the test the solutions that man, grappling with the major enigmas of his condition, must continue to invent if he is to survive".[22] In other words, to grasp the reality of individual and collective existence, to incorporate, organise or sometimes to dismiss it. Purveyor of meaning, purveyor of history, founder of an origin or warning of an end, the mythical narrative also serves to look into the future, to make decisions, even to found a society, to establish its rules, customs and as such, could never be reduced to something picturesque or imaginary. Perhaps therefore, we could consider any discourse that tries to answer some all too familiar questions such as "Who are we?", "Where do we come from?" or "Where are we going?", as mythical or having a mythical dimension. Space must not be afraid to turn to myths to find its foundations, its purposes and its responsibilities.

3.3 No fire in the sky without smoke on Earth

Space adventure begins on Earth. Instead of the wings of Icarus or Menippus, the descendents of Daedalus must resort to powerful engines, to blazing rockets: nothing about these astronautical technologies is reminiscent of the silent beat of the feather, the gentle swish of the air. To reach the sky, engineers, the modern-day Prometheus, were forced once again to steal the fire controlled by the gods! A power of this force cannot be used without causing pollution very similar to that sometimes associated with Earth-based industries.

Thus, with regard to the French law, the CSG, located in Kourou in French Guiana, is considered as a classified establishment and is subject to authorisation, considering environmental protection. According to the terms of the European Seveso directives (1982 and 1996), this qualification is aimed at "the prevention of major accidents which involve dangerous substances, informing the public of the risks involved and the limitation of their consequences for man and the environment, with a view to ensuring high levels of protection". At the CSG, nine sites are concerned by this classification, in particular the buildings used to store hydrogen and oxygen. In the event of an explosion, an area stretching fourteen kilometres from its source would be threatened. In order to respond to local concerns regarding space activities and their impact on human health and the quality of the environment near to the CSG, a permanent secretariat for the prevention of industrial pollution (SPPPI) was set up in August 1997.

Failure or destruction of the launch vehicle during launch operations constitutes the biggest risk of accident and pollution. The immediate protection of populations is therefore a major concern, both in the event of the self-destruction of the launch vehicle on its nominal trajectory (in other words, on the planned trajectory) and in the event of remotely controlled destruction. Even the lightest fragments can cause damage when they are carried by high-altitude winds. Finally, the passage of the launch vehicle upper stage over continents must be compatible with the safety objective of not claiming any victims.

The flights of the Ariane launch vehicles are subject to such constraints, based on fall out forecasts and modelling. These calculations were validated during the failure of the Ariane 501 flight in June 1996. However, control measurements are systematically taken during the preparation and performance phases of each launch. The CSG entrusts the management of a possible crisis during launch operations to a structure comprising four units: a management unit which manages the crisis, a communication unit responsible for internal and external information, an administrative unit which collects data and analyses the development of the situation, and finally a logistical support unit responsible for facilitating the work of the other three units.

However, the impact of launch activities on the environment is not limited to accidents and failure. Pollution also occurs during correct or nominal operation due to the type of fuels, solid or liquid, used. Recurring tensions between the Ukraine and Russia concerning use of the Baikonur base often concern pollution. The Ukrainians are asking the Russians for better guarantees in terms of environmental protection, an ecological demand that does not rule out financial disputes!

The Ariane 5 launcher was also the subject of criticism from ecologists in French Guiana. As a result of these claims, regulations in force and the policy implemented by CNES in terms of the environment, this reality and these risks are now taken seriously. Special attention is paid to the vast quantities of water poured onto the launch pad during launch, in order to cool the platform and weigh down the combustion cloud. In this way, most pollutants emitted fall out within the vicinity of the launch area. The remainder forms a cloud which stabilises at an altitude of 1500 m. Blown by air currents, alumina particles filled with acid scatter before falling back to the ground. Generally speaking, all "process" water (i.e. water produced by an industrial process including water from the various rainout processes of propellants or water retrieved in the launch area ducts) is analysed before being disposed of. Samples of surface and underground water taken from three locations (one upstream and two downstream of the facilities) are analysed every 6 months. Measurements are taken once a year on sediments and aquatic flora and fauna. In terms of the air, specific attention is paid to monitoring alumina fall out on avifauna, the mix of alumina and hydrochloric gas on flora in the form of throughfall or dust fall out on the urban areas of Kourou and Sinnamary.

Hydrochloric acid and alumina are produced by combustion of the powder acceleration stages and constitute specific risks. Hydrochloric acid (HCl) is toxic for humans and in solution can burn the skin. In gas form, it can lead to suffocation and breathing problems. Aluminium oxide (Al_2O_3) is hazardous not due to its chemical nature, but due to its dust state. Albeit rare, pulmonary fibrosis has been observed, in particular among minors. The aluminium ion is also phytotoxic. However, it should be remembered that Guyana soil is naturally rich in alumina, e.g. in the form of bauxite in the Kaw region.

Analysis plans and monitoring campaigns (during a launch, more than a 100 points are studied and 600 samples are analysed) produce the following results today. The impact of launches on air quality appears localised to within the immediate vicinity of the launch area; beyond this, fall out is low. Process water is rejected with characteristics which conform to regulations in force. In terms of surface, underground or river water, there is no notable physico-chemical variation, in relation to the initial state. Impacts are visible on avifauna near to the vicinity of the launch area, in particular on the outer feathers. However, the metabolism, analysed based on the observation of outer feathers, would appear to stay the same. There is no effect either on the behaviour of birds or on the quality of bird populations. Some species of aquatic fauna present higher aluminium concentrations. However, not all rivers are affected. In terms of vegetation, this is mostly affected within a radius of 250–500 m around the launch area. High concentrations of chloride and aluminium cause leaf necrosis. However, beyond this, the impact is less noticeable.

Experts admit that they have reached the limits of simulation and metrology studies and are now looking to develop a global approach to these phenomena, paying specific attention to interaction processes of the natural environment with all spot measurement variations. There is still not enough data; so to remedy the situation research is being conducted into biological indicators. The issue is therefore a long way from being resolved and there is enough pressure on space industry players for the development of less pollutant fuels to remain on the agenda without, sadly, the drawing up of an international regulation or agreement being currently conceivable.

To conclude, let's look at a surprising initiative implemented by NASA regarding the use of some vapours produced during shuttle launch preparations to produce crop fertilisers. These vapours, collected in particular when the tanks are filled, can react with potassium hydroxide to create potassium nitrate, e.g. a fertiliser readily available on the market which could be used to fertilise the orange trees at the Kennedy Space Center! Is "green space", "eco space" in our sights?

[18] The quotations are taken from book VIII of Ovid's *Metamorphoses*, in the translation by Frank Justus Miller. Cambridge: Cambridge University Press, 1984.
[19] In a letter sent on 12 August 1911, to engineer Boris Vorobiev.
[20] See www.nss.org and the magazine *Ad Astra*.
[21] See McCurdy, Howard E. *Space and the American Imagination*. Washington/London: Smithsonian Institution Press, 1997.
[22] See Lecourt, Dominique. *Prométhée, Faust, Frankenstein. Fondements imaginaires de l'éthique*. Paris: Synthélabo, 1996.

CHAPTER 4

4 Cloud riders

Forty years after man first set foot on the surface of a celestial body other than the Earth, the debate over the opportunity of sending humans into space remains topical. The place that humans currently occupy in space exploration and conquest is perhaps changing compared with what it has been until now. This change is both qualitative and quantitative. It is qualitative in that pioneering spirit is no longer what it was. Astronauts, cosmonauts and taikonauts continue to recruit from among the top aviation pilots although they are gradually being joined in space by engineers, scientists and doctors. The "right stuff" has certainly not disappeared but its makeup has altered. Without becoming banal, space has become humanised, if only due to the influence of women. Without doubt, when exploration of the planet Mars by a team of humans (when? no one today is bold enough to set a specific date), is no longer a dream but is actually being implemented and planned, will this "stuff" once again find its initial splendour, reminding us of the days of Gagarin, Glenn, Armstrong or Aldrin? However, between now and then, a quantitative change will probably have occurred, with a significant increase in the number of astronauts. The completion of the construction of the international space station in circumterrestrial orbit and its occupation has already played a part. What is to stop the development, even small-scale, of space tourism? For all that, however common the presence of humans in space might become, the questions "how" and "why" will continue to be asked.

4.1 The right stuff

Astronauts continue to inspire dreams. Through them and the flag they wear on their spacesuits, according to a well-known phenomenon of identification, citizens of an entire nation, the inhabitants of an entire planet, participate in the adventure of manned flights and, indirectly, track the path of the stars. The astronauts of the Apollo missions were thus surprised to hear men and women saying: "Thanks to you, We have been to the Moon". However, this dream, particularly since the Challenger and Columbia shuttle disasters, must come face to face with reality: close to 500 men and women have travelled into space since the first manned flight, but 22 have died in accidents, during a mission or during training. This figure is far from negligible. It is not surprising if public opinion questions the opportunity of pursuing such a risky and humanly expensive venture, adventure, as that of manned space flights. André Lebeau echoes this when he asks: "Does planetary exploration exceed the limits of what we can claim to achieve by adopting a strategy which refuses to accept, as part of the normal course of things, the sacrifice of human lives?"[23]

4.1.1 Man, an animal doomed to risk?

We have a strange relationship today with the notion of risk. On the one hand, we demand that public and government institutions (I speak essentially about most "western" societies) improve the management and control of risks to which individuals and groups are constantly and inevitably subjected, whilst on the other hand, we are ready to choose to behave in a certain way and to undertake high-risk activities (I am referring both to the use of drugs and to the practice of so-called "extreme sports"). This strange relationship may become paradoxical, even conflicting, when an individual dramatically involved in a high-risk activity expects and even demands that society help him/her out of a "tight spot". Various debates are taking place in Europe concerning the practice of mountain sports and the cost of rescue missions if these extreme sportsmen and women fall into danger. This delicate and thorny question also concerns the relationship between the private sphere and the public sphere: a relationship that cannot be settled once and for all; a relationship that scientific research and technological development cannot ignore. Manned flights and the sending of astronauts on long-term missions are no exception. They raise issues and specific questions that could capture the attention of the research community.

If we are to believe François Ewald, man is an animal doomed to risk. Is risk specific to humans? Does the simple fact of existing not lie at the root of most risks faced by all living beings? There are natural risks against which scientific progress can only ever offer partial or temporary protection, risks for which only biological chance, destiny or fatality may be summoned. However, technological, industrial and military accidents and professional misconduct were created by man "due to his inability to master all the elements of the systems he has designed, in his haste to apply on a large scale solutions or products that have not been tried and tested, through his failure, recklessness, violence or unreasonableness, etc." (Jean-Jacques Salomon). Effectively, if we look at it this way and consider the role played and the place occupied by technology, the human being is an animal singularly doomed to risk. But what is a risk?

The authors of *Logique de Port-Royal* (1662) wrote "The fear of harm must be proportional not only to its gravity but also to its probability". What these authors qualified as "fear of harm" is what is today known as risk. All decisions concerning a risk to be taken or to be avoided consider two inseparable yet distinct elements: on the one hand, objective facts concerning its possible occurrence, on the other hand, the more subjective point of view relating to chance of loss or gain associated with this decision. Major risk is thus qualified as a low probability occurrence that can however destabilise a given community, trigger a crisis and require the implementation of exceptional means. Risks associated with the use of atomic energy and nuclear technologies fall into this category. However, those associated with cars and aeroplanes do not. Even if taking an aeroplane presents a danger with serious consequences for individuals, this means of transport is considered with a low probability of accidents by most people. The risk associated with the use of the car is more difficult to determine: the probability of an accident is far from being negligible but the possible types of damage and injury are extensive.

What is the risk associated with rocket launch? Depending on the type of launch vehicle used, the probability of failure (most often due to explosion of the engine) is around $1\text{–}5 \times 10^{-2}$; in other words, the permitted or noted failure rate is 1–5%. The fixed objective for the Ariane rocket family is more specific: the risk or more precisely the probability of claiming a victim during a launch, no matter where in the world, must be lower than 10^{-7}. In practice, it is not always possible to adhere to this objective and concessions are granted to make the probability 10^{-6}.

This simple example provides a wealth of information. Until recently, the idea of failure was excluded in technical environments: the engineers' ideal focused on an objective to be reached, a success to be planned for. Thinking of failure would have seemed, to say the least, inconvenient, a waste of time and energy, a weakening of the ideology of progress and success. Moreover, how can the possibility of an accident be evoked when such a thing is defined as the unfortunate encounter of two causalities not properly controlled because of chance or human error? Do engineers, technicians and researchers not claim to control all causalities, to make them work for mankind? In other words, from a positive outlook that attempted to exclude all uncertainty, risk was considered as a danger to be eliminated and zero risk as an objective to be systematically achieved. The phrase commonly (and perhaps wrongly) attributed to Eugene F. Kranz, the famous NASA flight director, "Failure is not an option", was primarily the slogan of a generation of engineers....

Theoretically, taking risks is not contrary to ethics, morals or deontology. However, risks to be taken must be suitably assessed to limit negative consequences and to inform potential victims (individuals and institutions). Do space agencies and governments really think of this when they decide to launch a rocket that is to fly over several of the world's regions without the populations of these regions being necessarily informed? Whilst the risks of fall out are low, they are certainly not negligible. So, is it acceptable to take the running of a risk for another person who has not been prewarned for granted, even when we are talking about residents of a country with restricted access? What information and communication policy could be established to try to improve this kind of situation and to allow as many people as possible to benefit from progress in terms of risk assessment and management?

Risk assessment is now an essential part of the social, economic and political management of projects, whether it concerns the skill of engineers or the skill of programme managers and decision makers. Risk assessment is of particular interest to space agencies and public establishments such as CNES, as they must take responsibility for risks that manufacturers cannot or refuse to take in the domain of research and development. At the same time, when we are talking about human lives, it is not a question of taking more risks or as many as one individual can. Public opinion seems to accept the risks inherent to some exploits which are qualified as private and which are practiced by modern-day sportsmen and women and explorers. However, this same opinion is in theory against the risks taken by astronauts, even if they also are fully aware of the risks before engaging in their activities. Is this due to greater public identification with astronauts, to the public nature of space funding or to often heightened media coverage, although this latter characteristic has significantly changed since the first astronautical exploits? We must without doubt add to this the specific cultural behaviour and differences (e.g. in terms of the importance accorded to national heroes), in addition to the role of secrecy in the running of public affairs.

4.1.2 The end of zero risk

Times have changed. The people responsible today for developing and operating space launch vehicles now recognise that whilst a certain level of risk is inevitable, risks can be managed and reduced. They must abandon simplistic and determinist explanations of phenomena and the ideal of unlimited rational knowledge, to focus on an understanding and an apprehension of reality under the mode of complexity and the tangle of factors, most often using statistical calculations. The current approach has not therefore lost the resolute character which is inherent to the profession of engineering, but it takes into account the increasing complexity of both the systems in place and of sciences and technologies in general. For their part, sociologists are now interested in technological macrosystems, which play a central role in the uniformity of exchanges and unification processes all over the world. It is at the interface between the equipment factor (the artefact of the machine), the organisational factor (in particular the information network) and the human factor (e.g. hierarchical relations) that different types of accident occur, and precisely due to this confusion of technical and social, of machine and human.

So with whom therefore should the responsibility lie in the event of failure? No one can claim to be or be labelled an expert in all domains anymore. The complexity or even the opacity of systems eventually surpass their designers and lead to failure. Human error has become the main source of accidents.

As the myth of zero risk becomes a distant memory, engineers and societies need to manage the risks now recognised as being inherent to modern technological developments. How can this be achieved? Making a distinction between deliberate and unintentional risks, immediate and delayed risks, irreversible and non-irreversible risks, etc. is without doubt necessary to facilitate their informed acceptance by users of the technologies in question or potential victims. An object, a technology, an operation or a process are considered to be safe when the risks they can cause to their users are known and deemed acceptable by the latter. In other words, a risk is better accepted by those exposed to it the more they feel it is their choice whether to take the risk; to be able to possibly control it, and not to be subjected to it against their will. It is probably due to this reason that, during enquiries into risk perception, the people interviewed feared plane accidents more than road accidents. That said, the perception of risk, of its varying degrees of seriousness and, at the same time, the idea of security become dependent on an era, a group, a culture, and a scale of values and principles. Even at the time of the Universal Declaration of Human Rights (1948), human life did not have the same worth everywhere in the world. However, do not be fooled... applying the principle of informed consent in order to make the risks of our modern technologies acceptable is not without consequence or negative effect. The antidote to fear of risk is not always training or information. Knowledge can create fear and rejection as easily as ignorance. This poses a real dilemma, a challenge for communication managers, in terms of science and technologies.

This observation and this challenge being set, there is no question of settling with ensuring acceptance, to a greater or lesser degree, of the risks associated with technologies or, on a wider scale, associated with life within our modern societies. Engineers and politicians must attempt to improve control of risk factors, the causes of accidents and the processes likely to result in disasters. It has to be recognised that, unlike natural risks and disasters, human, and predominantly institutional liabilities, are evident in the technological domain. It is not a case of trying to revive mythology, warns Jean-Jacques Salomon, or of finding lowly scapegoats: "There is never inevitability, he writes, but a combination of circumstances in which appears first and foremost a lack in the vigilance of institutions, as opposed to an error in the interventions of individuals".[24] The expected and required vigilance, observed Patrick Lagadec in his work *La civilisation du risque* must on the contrary leave the future open: "A general effort in relation to social attitudes and behaviour; the civilisation of risk can only anticipate an open future if it continues to function according to the convention: after us, the deluge".[25]

4.1.3 Institutions and innovation

These thoughts can be applied within the domain of innovation. Innovation is probably the domain in which the intentionality of technical and techno-scientific projects is expressed most clearly. It is also the domain in which ethical questioning can be the most necessary. Bruno Latour writes: "The innovator's dilemma is well known; when he can choose, he does not have sufficient data; once he has the data, he no longer has any choices. At the start of his project, if he knows nothing yet of the reactions of the public, financial players, suppliers, colleagues and machines that he must bring together to make his project take shape, he can however very quickly completely change the nature of his plans to adapt them to their wishes. At the end of his project, he will finally have learned everything he should have known about the resistance of materials, the reliability of components, the quality of his subcontractors, the loyalty of his bankers, the passion of his customers, but it is too late to make any further changes to his plans which are now cast in bronze".[26] The ideal would be to be able to wait to know in order to act. However, it is usually necessary to act in order to gain knowledge. Luckily, between knowledge and ignorance, remains experience gained from both successes and failures. This experience should be gathered, analysed and preserved. Referring to the idea of falsification, embraced by Karl Popper and to epistemology, Latour notes that good experience is "that which puts the facts used to define the project to the test very early on", whereas bad experience is one from which we learn nothing. So, to contribute to progress, innovation must recognise failure and take risks.

Space activities are subject to the constraints of innovation although we probably should not exaggerate these. The idea according to which the conquest of space would be synonymous with the conquest of progress, accepted by public opinion and often driven by the media, is worth, without a shadow of doubt, being precisely defined and qualified. For his part, Roger Lesgards does not burden himself with as many precautions when disputing the idea according to which space projects could have crucial technological consequences: "Space, he writes in *Conquête spatiale et démocratie*, is essentially a user of technologies developed elsewhere, particularly in the military and astronautics sectors. This is as true for the information technology and electronics of satellites as for launch vehicle materials and propulsion. Of course, their use in space involves taking them to limits, adapting them to operate in a hostile environment, improving their reliability and the performance, 'spacialising' them ... but in so doing, this makes them unsuitable for use in industrial low-cost series production. The American government did its utmost to list these hypothetical 'consequences'. Result: a short and unconvincing list of several objects with a limited market. State-of-the-art, 'high turnover' industries owe practically nothing to space. There is very little technological transfer in their favour apart from a few effects on complex project management and quality control methods. However, military and aeronautical activities already have this type of advantage."[27]

André Lebeau's opinion is less categorical. He considers that "technological change is a global phenomenon to which space technology contributes but which, above all, is governed by external developments. [...] space policy can only consider technological development, not govern it".[28] This is saying that, without ignoring it completely, the ethical issue in terms of space, will only slightly concern the actual domain of technological innovation. Innovation will probably come from elsewhere, in particular from possible scientific discoveries.

4.1.4 From risk to caution

The theme of risk is also associated with that of safety, security and, consequently, the question of existence, alongside operational establishments and those responsible for monitoring and surveillance. In accordance with how the nuclear industry uses the term, safety is specific to the strictly technical domain, whilst the term security designates protection against external dangers and malice. These two approaches are not completely interchangeable. In terms of safety, it is possible to demand a real, practically technological, objectivity, whereas security is mostly associated with an era, a culture and other defining characteristics. If the same quality of safety can be demanded by the pilot of a prototype aircraft or the driver of a prototype car as by the operator of one of these vehicles once qualified, the level of security permitted by them is not the same. The difference stems from the difference that necessarily exists between innovation and common practice. The criteria involved in the switch from one to the other still needs to be defined.

In all cases, especially if the dream of zero risk has been abandoned and its illusory character recognised, it is still necessary to accept the possible existence of an indistinguishable or immeasurable risk. The cautious approach involves not stopping at what the knowledge of the expert or scientist can discover, analyse or master, but more in considering the limits of this knowledge. Caution rightly questions knowledge. Did Descartes himself not speak of a "hyperbolic doubt", of a hyperbolisation of scepticism? In other words, being cautious would consist in imagining what we do not know, of bearing in mind, when making a choice or a decision, both recognised knowledge and knowledge that remains unknown or is accepted as lacking. No one should claim to be skilled or an expert in any domain, without accepting the limits of their knowledge, skills and without knowing "the possible and the actual" (François Jacob). However, who among us is ready to doubt their own knowledge and their own logic?

4.1.5 Risks and remedies

Let us return to the field of space. There are many different types of risk run by living beings, in particular humans, during manned flights, depending on their origin (accidents, effects of zero gravity and of cosmic radiation), their impact (physiological, psychological), the term of their effects (short, medium, long) and the possibility of providing protection or solutions.

The shelved European Hermes shuttle programme provides an example of the constraints associated with the safety of crews, in terms of accidents. Following a number of technical, political and economical incidents, this programme was finally abandoned in July 1992. During previous years, in particular between 1987 and 1989, safety objectives were widely studied and discussed. The principle involved accepting a relatively high mission failure rate (in this case, up to 10^{-2} per mission), whilst implementing effective crew rescue means, in order to obtain a probability of crew loss of 10^{-4} per mission. In comparison, the probability of natural death per year is estimated at between 10^{-4} and 10^{-3}, depending on age, and the probability of the death of a fighter pilot or an extreme sports enthusiast can reach 2×10^{-2} per year. To achieve this objective, it was necessary to design an spacecraft with an actual "rescue cabin", effective both during launch and during atmospheric re-entry. Designing the cabin became the basis of the work for designing the European shuttle, during the first years of the programme. At the start of the 1990s, some unsolved technical problems associated with the existence of this rescue cabin still persisted. In addition, the spacecraft started to become too heavy for the performances of the new European Ariane 5 launcher and finally development costs threatened to largely exceed initial budgets. CNES and ESA thus decided to abandon the idea of a rescue cabin and moved on to the idea of ejectable seats, which altered the probability of crew loss to 10^{-3} per mission. While remaining in line with figures already given regarding the natural or accidental death of pilots, this new figure also corresponded to the estimated probability of crew loss following an impact with orbital debris.

The Hermes Safety Advisory Committee (or HESAC) followed up the Hermes programme in terms of human safety. This committee was particularly involved in the organisation of the programme, work methods, technical choices and risk analysis. It estimated that an objective of 10^{-3} per mission, associated with outstanding technical work, would result in better safety than an objective of 10^{-4} associated with insufficiently rigorous methods.

Surviving an accident in Space. American and Russian philosophies

When the very first manned flights took place with Y. Gagarin and J. Glenn, as part of the Apollo programme, the Russians and the Americans had a very similar approach to crew safety. Oddly enough, the Americans fundamentally modified their philosophy with the shuttle programme. However, after the Challenger disaster in 1986, NASA tried to return to the former concept without having to make too many changes to its flight vehicle, which would require improving the reliability of all of its systems and result in expensive operating costs.

At the start of the programme, the Americans and Russians fixed relatively high mission failure rates of approximately 10^{-2}, rates accepted by our societies for high-risk professions, such as the army. This was one of a number of reasons for the looking to military pilots when recruiting astronauts and cosmonauts. Three major principles were applied: (1) the safety of people on the ground before that of the crew, (2) crew solidarity: rescues attempts are aimed at the entire crew as a unit and not one particular individual, (3) safety approach from start to finish, using procedures and means specific to each flight phase, including on the launch pad and in the ascent phase.

The Russians succeeded in saving their crews twice. The first time was on 5 April 1975, 3 months prior to the joint Apollo-Soyouz flight. Following a separation problem between the second and third stages, the Soyouz 18-1 capsule performed an emergency separation from the launch vehicle, bringing the two cosmonauts V. G. Lazarev and O. G. Makarov back to Earth. These two men had to endure decelerations of approximately 20 g for 20–30 sec during atmospheric re-entry. The second occurrence took place on 26 September 1983, just over a year after the maiden flight of Jean-Loup Chrétien. Fire broke out on the launch pad 90 sec before scheduled blast-off of the Soyouz T 10-1 in which the two cosmonauts V. G. Titov and G. M. Strekalov were installed. They activated the launch escape system which fired with an acceleration reaching 17 g, at an altitude of close to one kilometre, and the descent module dropped under parachute 2.5 km from the launch pad on which the launch vehicle had exploded.

Did the success beyond all hope of the Apollo programme (Neil Armstrong and Buzz Aldrin were only given one chance in two of returning from the lunar

(*Continued*)

expedition; the dramatic yet successful return to Earth of the Apollo 13 Moon mission crew safe and sound, a mission aborted due to the explosion of a service module tank) and a subsequent boost in confidence, lead NASA to reconsider its philosophy? The fact remains that with the shuttle system, after the first four orbital flights carried out with a two-astronaut crew equipped with ejectable seats operable at Mach 3, the philosophy reached a total standstill over the possibility of rescue during the solid-propellant stage propulsion phase: 120 sec during which astronauts are at the entire mercy of technology, with consequences as we know for the Challenger team.

True to their tradition, the Russians stuck with their initial philosophy. Why change a system that works? You may argue that the Russians did not build other manned mission launch systems but that would be wrong. With the ephemeral Buran, the "almost identical" copy of the American shuttle, the Russians adhered to the philosophy they strongly believed in of safety from start to finish, by transferring all propulsion to the Energia rocket, allowing the space aircraft of the launch vehicle to separate with less risk, by refraining from the use of solid propellant propulsion, by providing Mach 3+ ejector seats and by automatically controlling the entire re-entry and landing phase, unlike the shuttle. Buran only flew one demonstration flight during which it performed a text-book landing. There was no pilot on board the aircraft; clearly the most effective way to eliminate all human risk for still very risky maiden flights.

Alain de Leffe and Jean-Jacques Favier.[29]

The absence of gravity remains, with the type of spacecraft currently used and envisaged, an essential element of life in orbit, probably one of its most spectacular characteristics and, at the same time, the source of many types of problem and risk to astronauts. Within the limit of stays completed to date, the main dangers are the phenomena of muscular atrophy and calcium loss which may be crippling, if they are not pre-empted by daily physical exercise (cycling, running on treadmill) which ensures a vigorous mechanical solicitation of the bones and muscles.

Cardio-vascular adaptation phenomena, associated with the movement of fluids to the *thoracocephalic* region, follow the disappearance of hydrostatic pressure during weightlessness; they run their course in several tens of hours without major handicap. The opposite adaptation, during return to gravity, has been improved through the use of equipment such as Low Blow Negative Pressure which subjects the lower limbs to a local negative pressure to cause the heart to function as closely as possible as it would on the ground: in other words, to pump blood from the legs to the upper body.

Finally, it is appropriate to mention the appearance of sensorial disorders, when visual and propioceptive information ceases to match information from the otolith system. The establishing of a new set of references leads to a phase of illusions and disorientation that can lead to space sickness. The intensity and duration of incapacity caused by space sickness vary from individual to individual.

Experts consider radiations to be the main restriction to long journeys into interplanetary space. In effect, beyond the Earth's atmosphere, there are many different types of primary and secondary radiation (i.e. created by the passage of primary rays through the structural materials of space craft), solar and cosmic in origin, all with significant fluctuations. Generally speaking, on-board hardware and equipment exposed to these radiations suffer damage that can result in the loss of system functionality, due to rapid ageing, background noises or false commands. Malfunctions of this type can potentially place crews in danger and therefore require the implementation of protective structures and surveillance procedures.

The dangers faced by living beings exposed to radiation are obviously significant. In the face of radiobiological risks, they are characterised by a major sensitivity to radiation and by a high repair capacity. In effect, with the exception of cerebral tissue and excluding massive radiation exposure, immediate damage can disappear in a few months due to the production of new cells. However, delayed effects also exist, which can be teratogenic, carcinogenic or genetic and astronauts will need specific medical follow up. On the other hand, immediate effects, which would result in a reduction of the capacities of crew members, and even in their death, are very unlikely or only possible during exceptional solar eruptions. It is should be noted that sensitivity to radiation is multifactorial: it simultaneously depends on the nature and status of the cells concerned, their physical and chemical environment, the intensity and time of exposure and finally, the nature of the radiation, the gender of the astronauts (women are more affected than men).

Permissible exposure limits for the entire career of an astronaut were suggested. The probability of contracting cancer being reduced along with the estimated lifespan following exposure to radiation, the acceptable doses were calculated to give a mortality risk of 3% for the entire career. The ALARA (As Low As Reasonably Achievable) principle is applied during each mission, particularly when working outside the space craft or during solar eruptions; in the latter case, in effect, the maximum monthly doses can be exceeded in a matter of hours for certain critical organs and immediate effects are probable.

As far as lunar and Mars missions are concerned, in other words, in the absence of Earth's protection, the level of exposure to radiation is currently a matter under discussion. Until now, the importance granted to levels of exposure to radiation was such that only crews of a mature age and without past exposure to radiation were considered, to avoid exceeding the maximum career doses. Along with the essential protection in view of solar eruptions, it is at least accepted that the radiobiological risks are extremely variable, that they depend on the actual mission phase, either travelling, or in station. There is still a long way to go and many risks still need to be assessed before a decision can be made regarding the feasibility of long missions.

Manned flights often provide opportunities for conducting scientific work on humans, with teams taking on the simultaneous roles of experimenters and guinea pigs. It is therefore advisable to adhere to a real deontology in terms of human scientific research. The idea of informed consent could even be applied to manned missions as a whole and to the exposure to risk that they constitute.

At first, astronauts and cosmonauts flew solo into space (Voskok, Mercury, Soyouz), then as part of a national team, into Earth orbit or during lunar exploration missions. Missions have progressively got longer, some even lasting over a year (Saliout, Mir) and crews have become mixed, international, intercultural and interracial. As logical as it may seem, this change brings with it new drawbacks and new risks, including the allocation and management of responsibilities, management of authority and compliance, personal and collective management of disputes (aggressiveness, etc.) and management of the personal sphere (leisure activities, etc.). The psychological and social dimension also plays an increasingly significant role in the preparation and performance of missions. We are without doubt no longer in the awareness or experimental stage: in 1995, NASA gave its astronauts their first Russian and Japanese lessons. The previous year, it had offered Russian cosmonauts involved in the Mir-Shuttle missions, lessons in American culture. The issue is still present: studies have shown that the behaviour of airline transport pilots depends on their nationality, their grade and their gender. Of course, airline company crews are formed for each flight, whereas, when it comes to carrying out their mission, astronauts and cosmonauts have already worked together for many months or even years. However, even this is not enough to erase cultural differences and therefore managing them remains an issue. Initiatives have already been taken in this domain: NASA has drawn up several regulations applicable to individuals on board the space shuttles, regardless of their nationality, concerning the possession of personal objects, the authority of the pilot-in-command, the chain of command, non-compliance with regulations potentially resulting in fines and even imprisonment, etc. Along similar lines, an agreement pertaining to *Cooperation on the Civil International Space Station* (1998) was signed between the various States and organisations involved in this programme. This text constituted a new chapter in the history of manned flights and international cooperation in more ways than one. According to the future planned for this type of mission, it will probably have to be added to so that the rights and duties of different crew members may be determined, insofar as their missions, tasks and occupations are going to become increasingly diversified.

4.2 The astronaut, *factotum* of space?

"Astronauts are no longer just pilots", wrote Gabriel Lafferranderie in 1993, "they are also scientists, astronomers, doctors, engineers and journalists and in the future gardeners, minors and eventually salesmen. They help to conduct industrial experiments. These 'odd-jobs men' live in a confined space, permanently observed and listened to by the ground crew, which will send them their work orders, will wake them up or ask them to sleep, to do physical exercises, or to undergo medical checks, to perform in space actions which do not come naturally".[30] The portrait that Lafferranderie paints of yesterday's, today's and maybe even tomorrow's astronaut is spot on: since the first flight of Yuri Gagarin, the astronaut has apparently become a 'jack of all trades'. How can this change from pilot to *factotum*, to a person who does everything and anything, as evoked by Lafferranderie, be explained? The most simple response would seem to be: over 50 years and through its 500 members who have torn themselves away from Earth's gravity, mankind is just learning how to live in space. From reading space literature, it would seem difficult to live in space, without working there!

Live and work: the expression returns like a leitmotif in a number of works that discuss the theme of mankind in space. These include publications for a non-specialist readership, historical works and directives concerning the selection of astronauts. The designation of the term "work" covers a wide scope: first and foremost, the tasks of daily life on board an orbital station or a space shuttle; then the carrying out of scientific experiments in the domains of physics, chemistry, biology, medicine, astronomy and Earth observation, including for strategic purposes; finally, the construction of a space station and, if necessary, the retrieval and repair of satellites. All being subject, as Lafferranderie reminds us, to ground control. In other words, the men and women who live and work in space, especially around the Earth, have very little time to sit gazing out of the portholes of their vessels, to contemplate their homeland Earth or the stars, to dream of *terrae incognitae*, of worlds still unknown and awaiting discovery.

As if to make their status even more difficult to define, in addition to the many tasks which astronauts have to perform there is their evolution throughout the (concise) history of space. There are no shortage of illustrations. We simply need to evoke here the method of recruiting astronauts: their corps, primarily national and made up of men (with the sole exception of the Russians), military personnel and pilots, are gradually embracing the inclusion of women, scientists, engineers, doctors and, finally, an international composition. Each member is obviously not asked to perform identical tasks.

It is also interesting to note that Skylab, a structure designed based on the Apollo programme and sent by NASA into Earth orbit in 1973 and 1974, was considered to be a workshop and not a space station, even though three crews of three men lived and worked there! The distinction was introduced to announce the future installation of a permanent and larger structure, as it happens with the current International Space Station or ISS.

The meaning of "live and work" has therefore changed referring to space. Whilst Icarus the explorer has not been totally pushed aside, the importance of Daedalus the engineer is heightened. His workshop must not however take up the entire spacecraft, as had long been the case, in the Skylab, Saliout or Mir stations. Even so, the life of astronauts is today still organised around work. In work, I confidently include the physical exercises which men and women who stay in space for several weeks or months are obliged to do. The designers of current and future space structures (stations or spacecraft) are focusing in particular on "living volumes" whether collective or individual. It will probably be necessary to make a clearer distinction between life and work, by introducing, among other things, leisure activities. I am not even thinking about space tourism, which signifies a division of work and rest between several people or groups of people, but simply of crews who have been sent into space on long missions, e.g. to Mars: their leisure time, as much as their work, would need to be thoroughly planned and adapted to being further from Earth than has ever before been attempted.

4.3 Is there room for advertising in space?

If, in the modern space adventure, it seems ultimately perfectly normal that Icarus the explorer cannot continue without Daedalus the engineer, the increasing presence of or the seat simply demanded by Hermes, the god of travellers, roads and crossroads, seems much more problematic, as does the matter of commerce. That a telecommunication satellite can take the place of an exploration probe on board a launch vehicle, even in circumterrestrial orbit, thus providing a commercial service is not in the least bit surprising or even appalling. It is simply an example of the way in which mankind manages (and has always managed) the transition from the exploration phase to the utilisation phase and then on to the commercialisation stage. With that which can or must accompany this in terms of politics and legal matters, social and cultural matters, with its share of fortunate and unfortunate consequences, Hermes, the god of trade, is also the god of thieves! Yet, that which can appear normal and evident in terms of satellites, orbit and the allocation of frequencies (not without the help, as I have said before, of regulatory bodies and procedures), is not so when it comes to astronauts. Two commercial offers, the marketing of the activities of astronauts and space tourism, currently illustrate and form the focal point of debates on the subject.

The descendants of Hermes, by whom I mean the trade sector, did not wait until the 21 st century to take an interest in space and to ask themselves how much money it could make them, excluding the remote sensing market, telecommunications and global positioning. Cosmic marketing took its first steps at the same time as the American astronauts first stepped onto the Moon, when their image became associated with that of the famous brand of watches Omega. It was apparently the Israeli brand of dairy products Tnuva, which inaugurated the use of space advertising, thanks to the Russian station Mir, in 1997. Kodak followed in Tnuva's footsteps with the international space station. In the meantime, Pizza Hut had painted its logo on the Proton rocket, launched in July 2000. There were also several other stagings, such as that of a cosmonaut with a stick of glue at zero gravity. Is there a threat that spacecraft will be painted in the colours of any food company with an advertising budget high enough to provide financial support to space? Will the future spacesuits of astronauts look like those of Formula 1 drivers, plastered in advertisements? Those in charge of the space agencies say no but some, such as ESA, have already enlisted the services of communication companies to work on the possibilities, terms and conditions of marketing some of the activities and equipment, on board the ISS.

Depending on their native country and their culture, astronauts have reacted differently to this marketing trend that could be seen as a kind of inevitability. Due to the economic climate in their country, the Russians have been only too happy to accept advertising contract proposals; the Americans were not approached due to NASA regulations, and the Europeans wanted to take up the issue, base their responses to these marketing requests on ethical thinking about the meaning of their mission.

4.3.1 European astronauts confronted with commercialisation

In 1998, the member States of ESA decided to group all of their astronauts within the European Astronaut Corps (EAC), created in 1990 and located in Cologne, in Germany. In 2002, members of the EAC drew up their own charter, in order to specify the vision, the mission and the values common to them all.

The Charter of the European Astronaut Corps

Our vision:

Shaping and sharing human space exploration through unity and diversity

Our mission:

We **shape Space** by bringing our European values into preparing, supporting and conducting space flights to advance peaceful human exploration.

We **share Space** with the people of Europe by communicating our vision and goals, our experience, and the results of our missions.

Our values:

Sapientia: we believe that Human Space Exploration is a wise choice by and for humankind. **Sapientia**, as a truly European and our central value, reflects our commitment to pursue our goals for the advancement of humanity.

Populus: we put people first – in a twofold way. The purpose of our missions is to pave the way for a better future for people on Earth. **Populus** means that we show respect for the people with whom we work together; that we value their opinions, praise their work and compliment them for their support.

Audacia: we acknowledge that Spaceflight is a dangerous endeavour. While accepting the risks inherently involved in space travel we work to minimise these risks whenever we can within reason. **Audacia** reminds us that the rewards will be unparalleled if we succeed.

Cultura: we continue the exploration started by our ancestors. Conscious of our history and traditions, we expand exploration into space, passing on our cultural heritage to future generations.

Exploration: we value exploration as an opportunity to discover, to learn and, ultimately, to grow. We are convinced that humankind must embrace the challenge of peaceful human space exploration. We, the European Astronauts, are willing to take the next step.

This charter accompanied and served as a springboard for the thoughts of the European astronauts themselves, from the year 2000, concerning the possibility and opportunity of commercialising their activities in space. Based on these thoughts and following the work of a group of experts brought together by the EAC, the ESA management team introduced a decision-making procedure in order to deal with the commercial soliciting of astronauts. In 2003, it also set up an *ad hoc* committee of experts, the Astronauts Commercial Activity Board

(ACAB), later to become the Independent Commercial Activity Board (ICAB). The ICAB charter specifies the mission of its members: (1) to intervene on ethical questions and issues raised by the commercial exploitation of the image of ESA astronauts, in order to assess, from a humanist perspective, the legitimacy (in terms of opportunity, acceptability and feasibility) of proposals made to ESA in this domain; (2) to develop expert skills, a vigil capacity and drive within the scope of a mission, without reducing the mission to that of an emergency response unit; (3) to improve the image of ESA, its policy and its choices, within European societies. As part of its responsibilities, the ICAB also approves or rejects proposals involving the use of the personal image of ESA astronauts in marketing activities associated with the ISS, in accordance with the terms and conditions set out in the documents *Policy on the involvement of ESA staff in ISS Commercial Activities and Evaluation Procedure for commercial offers using the International Space Station*. It is reminded that "the personal image is not only the visual representation or the name of an individual, but also how he is seen, summarised or perceived by himself, his family, his acquaintances, the media, his peers and the community at large", that "any person has an absolute right over his/her own image and its utilisation. This allows him/her to oppose any exposition, reproduction and dissemination in any support of his/her own image, without previous authorisation". It has to be known that the members of the ICAB, appointed by the director general of ESA, have very rarely all come together in recent years. An illustration or proof that once the effects of the announcement have past, there remain few marketing opportunities associated with the ISS, in particular those which could use images or the personality of the astronauts. This does not mean that the questions asked at this time were not important.

4.3.2 Newcomers to space

What in fact do we know about what the public think about the image of astronauts, whether it is promoted, expected or dismissed by public opinion? And what do we know about how astronauts see themselves and the image they want to portray to the public? The possibility of the marketability of astronauts depends on their personality, how they are perceived by the public, how well known they are, and their sense of humour. It is however important to remember that the use of their image may belittle their missions and be detrimental to the organisation to which they belong. What differences should be drawn and respected between the images of astronauts and those of the other public figures so dear to our modern societies, their opinions and their passions: researchers and political personalities, show business stars? What about sportsmen and other "conquerors of the futile"? Without belittling their efforts, these men and women are perceived as heroes within our societies, but also entertainers; otherwise, how could we accept that they be plastered, to use the term that first comes to mind in relation to marketing, in advertising slogans? No one has ever really taken advertising seriously, apart from in terms of the amounts of money involved. Yet it still plays an important role within what we today call mass media and show business. Advertising has invaded television studios and sports stadiums on a massive scale but remains discrete or even inexistent in other leisure or cultural venues such as theatres and scientific research environments (e.g. Nobel prizes) or those of philosophical and religious movements. Is this due to lack of commercial interest in these environments or out of respect for function and what they represent? This statement would be difficult to support, as sport is a vehicle for essential values, especially for young people.

Be that as it may, astronauts neither are, nor are perceived to be, entertainers. Their exploits may incite admiration, fear and sometimes even a smile, but they cannot be categorised as "conquerors of the futile". Their venture must constitute an achievement for mankind, which I usually refer to as an "accomplished feat". Just as rugby players or fans talk about a converted try, a feat is accomplished when the venture in question is not only concerned with individual performance, but also with the good of mankind and its future and well-being. The space venture offers many examples. In terms of manned flights, it seems that a stay in circumterrestrial orbit is in the process of becoming an accomplished feat: the proof will be the arrival of the first space tourists. However, I believe we are allowed to say that Armstrong's "small step" onto the Moon cannot be considered an accomplished feat as twenty years after the success of the Apollo programme, no more humans have returned to the Moon. I am not saying that these missions have brought nothing to mankind; I only believe that they could have brought so much more and that they thus have an unaccomplished nature. I am well aware that this subject would make an interesting and worthwhile topic for debate.

In any case, to impose this value of accomplished feat on space venture is to return to the perspectives given to space by international treaties, out of their concern for mankind and its historic progress, out of respect for common heritage and the province that space constitutes, according to its dual definition. Astronauts, the acting figures in space exploration, are among those initially responsible for this concern and this respect. This is not simply wishful thinking: the media's attitude to astronauts, at least up till now, has shown as much respect, if not more so, for the individuals themselves in the name of this responsibility they bear (whether public opinion is aware of this or not), as for their actual activities. I think this could also be true of other domains such as armies, politics, philosophy or religion. Although it is common to stage their events and the practices that punctuate them, it would seem more difficult, even inconceivable, to see their members presenting on their official uniform advertising slogans. While the images of Earth taken by the American astronauts during the Apollo missions have been widely used, while films inspired by manned flights have been made (from *Apollo XIII* to *Space Cowboys*), while advertisements have used the theme of moon walking, without anyone complaining, the astronauts themselves have rarely been showcased.

Having summarised these limits and constraints, the possible marketing of astronauts' activities offers the chance to concretely raise the question about the valorisation of space activities. By this I mean their added value, what these activities can receive or offer.

Thus, the realisation in space of a scientific programme more often than not leads those in charge to contact an industrial company to develop specific products or instruments. This company may consider using this collaboration to promote itself. To reply to such a request, it is necessary to examine the proposed use, the message transmitted, and the content itself of the experiment concerned. Astronautical and aeronautical reviews are already filled with pages of advertising based on the images of launch vehicles, satellites or the ISS in order to extol the great technological skill of the manufacturer which assisted with these space programmes. Although I do not believe that this discredits these space programmes, are there not even so limits? When *Benetton* announced it was to begin an "experiment" on board the ISS in order to test the calorific quality of a fabric, was the term "experiment" appropriate? Do the conditions and technologies of space provide or receive any specific added value?

Furthermore, a marketing company can ask astronauts to promote its products from space, without directly becoming part of the space venture, simply only wanting to use the image of the astronaut. At first sight, the greatest caution is thus required: there is a risk that the astronaut, floating in his spacecraft, will serve the sole purpose of a space placard, gaining nothing from the exercise other than financial remuneration. A specific case could be when an astronaut would wear EAC brand products. The body of astronauts would therefore be promoting its own image, its own "philosophy" and in this case, would not serve as a silent middleman.

We have to acknowledge that the situations with which astronauts are faced, individually or collectively, are as varied as they are complex and are not expected to get easier. Is it up to them to adapt, mandatorily if necessary, to this complexity and its developments? At the risk of appearing conservative, I would be tempted to say no. I do not believe that astronauts should add the role of "sales representative" to their current qualifications and skills. If industrial or commercial companies believe that using the image of a man or a woman in space can provide added value for the promotion of their activities and their products, let them ask someone whose job it actually already is to work in space to help, and if that gets them nowhere, let them find someone else to go into space. Space tourists already exist. Why not consider training space actors to use in the shooting of clips and films? Establishing the terms and conditions of their access to space stations and spacecraft would not be that difficult. In other words, rather than see the appearance of "commercial astronauts", why not in due course employ "space actors" who could join "space tourists"? That would be a way of acknowledging the changes taking place or those forthcoming, within the space venture, concerning its progressive commercialisation in particular. That said, I do not deny these "newcomers" the qualifying title of astronaut to which the law associates rights and obligations. I only seek to preserve the symbolic and real responsibility of an astronaut, even if it is sometimes not clear: that of assuring and dealing with the ultimate progress of mankind, in its propensity for exploration and expansion. This tendency is expressed in domains as diverse as science and technology, general creativity and arts. However, one of the first domains, within the symbolic and geographical scheme of things, is without doubt that of space. Although this space expansion is currently limited to circumterrestrial orbit, we can decide to leave open the opportunity of going further, at a later date, whether this means returning to the Moon or reaching Mars. We will therefore need pioneers, men and women with "the right stuff", and not only the stuff of tourists or entertainers. We are without doubt still a long way from that and this is precisely what makes this period so difficult to manage in human terms. However, may I reiterate that we can decide to preserve, within the identity of the astronaut and in that of the researcher or the artist, something of the dignity of the explorer and, finally, of human beings themselves. As the ESA document that sets out the mission of the ICAB specifies, "the ethical rules may have to be perceived as time and culture dependent phenomena, i.e. 'living documents', with room for dynamic developments in the future".

4.4 When tourists rub shoulders with astronauts...

On 30 September 2009, Guy Laliberté became the seventh space tourist. This number is symbolic in space adventure: the first American astronauts, those of the "right stuff", were seven in number and the crews of the Challenger and Columbia shuttles, respectively destroyed in 1986 and 2003, included seven members... What exactly is space tourism? First and foremost, it is a commercial activity that offers its customers the chance to travel in space. Some authors readily include in the term those activities that only provide an indirect experience of space, such as zero-G weightless flights. In other words, space tourism does not so much offer a means of transport but the possibility of experiencing a situation and feeling sensations specific to the space domain including zero gravity and the viewing the Earth from a very high altitude. This activity is not about usefulness (at least for those who practice it), or scientific objectives (or only in a limited way). It is primarily about the pursuit of pleasure, leisure and perhaps even a type of spiritual experience (in the widest and least religious sense of the term).

Up until now, the only form of space tourism in the strict sense of the term has been a stay in orbit, on board the international space station. The Russian federal space agency is the only agency to provide the means to get there, on board a Soyouz spacecraft, after training lasting several months. The American agency Space Adventures is the one responsible for organising the flight for which the "ticket" will set you back between twenty and thirty million dollars.

In addition, the award of the prestigious Ansari X Prize to Burt Rutan's rocket plane, SpaceShipOne, on 4 October 2004, could now mean a suborbital flight experience in store for future space tourists. On 27 September 2004, Richard Branson and Burt Rutan signed an agreement, at the term of which, Rutan's company, Mojave Aerospace Adventure, would provide Virgin Galactic with five SpaceShipTwo space vehicles able to fly two pilots and six passengers to an altitude of 110 km. An estimated 200,000 US dollars will buy you a flight lasting three and a half hours, with three to four minutes of weightlessness. Over two hundred test flights are scheduled with the aim of obtaining the approval of the Federal Aviation Administration, without which these commercial flights could not get off the ground, particularly with regards to the insurance companies. Virgin Galactic estimates a one year gap between SpaceShipTwo's inaugural flight and the first commercial flight.

Other companies have plans to join this market which has estimated revenues of 350 million dollars, by mid 2010. Some companies (including Rocketplane, XCOR and EADS) are working on rocket-planes like Virgin Galactic, whilst others are working on capsules, with return by parachute (Blue Origin, Orbspace and TGV Rocket). However, the idea of orbital hotels such as that imagined by Robert Bigelow, still belongs to the world of science-fiction.

In more ways than one, space tourism raises several types of issue: the speed at which the astronaut-explorer sees himself not followed but accompanied by the tourist, the risks run by these new space travellers and the levels of insurance required, the legal challenges to take up, etc. On many levels this new activity seems to be pushing the space sector to its limits.

The first and most important limit is that of safety. The figures speak for themselves. To date, 4% of astronauts who have travelled into space have died. Of course, this in no way means that the astronautical engineers and space agency managers take the lives of crew members lightly, but that the domain of manned flights is still at an experimental stage, at the limits of technological skill. By way of comparison, it is important to remember that, of the thirty SR-71 planes (an American spy plane capable of flying at Mach 3.5 at an altitude of 26,000 m), twelve were accidentally destroyed, luckily with no human fatalities. One author points out that the effect of the scale of the production or the market needs to be taken seriously, comparing the handful of SR-71s or the production of commercial rocket-planes, which he estimated at one aircraft built per year, with that of Airbus 320 planes manufactured at the rate of one plane per day. So, how is it possible to consider the current forecasts of one accident in ten thousand or fifty thousand commercial suborbital flights as anything other than optimistic? It is not a question of asking whether such an accident will occur, but when, with all the consequences this will have on the development and continuation of a commercial company.

To those who, from this situation, conclude that the risks inherent to space tourism belong explicitly to the field of space as its raison d'être and cannot be compared with the risks involved in flying or using an aircraft, some would reply that, at least with regards to suborbital flights, most operations fall within the air traffic domain. Otherwise, why would it be necessary to obtain authorisation from the Federal Aviation Administration? Could we therefore envisage a dual jurisdiction both in ethical and legal terms: air traffic first and then space? In this case, a boundary would need to be established between the two domains. This has been attempted before but never seen through to the end. Some people support the theoretical altitude of a 100 km whilst others note that 30–40 km is already the flight limit for aircraft. This a second important limit and vulnerability of space tourism.

The third limit also concerns regulations and the legal implication for countries, in particular through the notion of launch State. To what extent must this fundamental principle of space law be applied, within the framework of a privately-owned company? Do States need to protect themselves against excessive involvement? This question may be even more important as the dangers faced and the damage caused by this activity are not *a priori* limited to only pilots and passengers but may also affect third parties. It is important to remember the constraints to be adhered to or to be imposed in terms of the production and management of circumterrestrial space debris (see below) or the protection of historic sites. If it becomes possible one day to visit the Moon landing sites of the Apollo missions, what rules will need to be drawn up and complied with? Although space law exists first and foremost at international level, it is also transposed into national legislations which are inter-harmonised. Should specific attention not be paid to the possible emergence of an equivalent, in space tourism, of maritime flags of convenience? Space must find its own happy medium, between too many and too few legal and regulatory requirements.

A fourth limit more directly concerns those who pay or who will pay for their ticket to space. Should they continue to be called tourists? They themselves find this title a little insulting and prefer the title private space traveller. In addition, the regulations established in 2002 to manage the international space station make a distinction between "professional astronauts" and "spaceflight participants". Is that not a diplomatic way to apply to everyone the rights and obligations, the honour and responsibility of being the representatives of mankind, at whatever cost?

The still-limited number of tourists appearing among the ranks of cloud riders is only a meagre illustration of an altering perception of space and of the Earth, one in which each of us can now play a part. Together with these heterogeneous crews arriving at the international space station, we can see on a daily basis, with hope or horror, that we humans are all part of the same crew on board a single vessel: the spaceship Earth.

[23] Lebeau, André. *L'Espace. Les enjeux et les mythes*. Paris: Hachette, 1998: 185.
[24] Salomon, Jean-Jacques. "Le risque technologique et l'aventure spatiale". *Zénon*. 2 (1997): 38.
[25] Lagadec, Patrick. *La civilisation du risque. Catastrophes technologiques et responsabilité sociale*. Paris: Seuil, 1981: 221–222.
[26] Latour, Bruno. "Comment évaluer l'innovation?" *La Recherche*. 314 (1998): 85.
[27] Lesgards, Roger. *Conquête spatiale et démocratie*. Paris: Presses de Sciences Po, 1998: 24–25.
[28] Lebeau, André. *L'Espace. Les enjeux et les mythes*. 13–14.
[29] Quoted in Arnould, Jacques. *La seconde chance d'Icare*. 199–200.
[30] Lafferranderie, Gabriel. "Espace juridique et juridiction de l'espace." *L'Homme dans l'espace*. Ed. Alain Esterle. Paris: Presses Universitaires de France, 1993: 255.

CHAPTER 5

5 The Spaceship Earth

At the end of the 19th century in Russia, Konstantin Tsiolkowsky imagined the precursors of our rockets and wrote the following, as previously quoted: "Earth is the cradle of Humanity; but mankind cannot stay in the cradle forever". Did the passing of the 20th century and the start of the 21st century prove him right? There is nothing to confirm this. Certainly, his compatriots, as Soviets, became the firsts to send a human being into circumterrestrial space and the Americans planted their flag on the surface of the Moon. Through its representatives, the cosmonauts, astronauts and taikonauts, mankind had effectively ventured out of its cradle. But for how long? Several days, several months, at the most. How far? From the Earth to the Moon, at best. Still a long way off, perhaps even unreachable, is the moment when mankind, reduced to a group of pioneers, will be able to definitively leave its terrestrial cradle, to pursue its singular odyssey elsewhere. The Earth, once the cradle of mankind, would become the spaceship with humans making up part of the crew, the boat which they must salute, and also the prison from which they might never be able to escape. Astronomer Fred Hoyle had, it seemed, predicted this contradiction, this opposition to Tsiolkowsky's prophecy, even before the success of Sputnik and Gagarin's inaugural flight. In 1948, he wrote: "Once a photograph of the Earth taken from the outside if available, a new idea as powerful as any in history will be set loose".[31] What idea is he referring to? Maybe that the drama of mankind would now take place on a planetary and not cosmic stage, that of the Earth, now in its globality observed, understood, even mastered.

5.1 The Earth delivered to your doorstep

"To be lifted to the summit of the World Trade Centre, explained French sociologist Michel de Certeau, is to be lifted out of the city's grasp [...] When one goes up there, he leaves behind the mass that carries off and mixes up in itself any identity of authors and spectators. An Icarus flying above these waters, he can ignore the devices of Daedalus in mobile and endless labyrinths far below. His elevation transfigures him into a voyeur. It puts him at a distance. It transforms the bewitching world by which one was 'possessed' into a text that lies before one's eyes. It allows one to read it, to be a Solar Eye, looking down like a god. The exaltation of a scopic and Gnostic drive: the fiction of knowledge is related to this lust to be a viewpoint and nothing more."[32] By imagining they could fly, our ancestors thought they would be able to look down like the gods and by fulfilling their dreams, they discovered not only that they could see, but that they could hear and even speak like the gods. From space, and due to the existence of space, everything seems accessible to us, whether it is a matter of seeing and hearing or of making ourselves seen and heard.

Philosophers have not missed the opportunity to highlight the possible danger regarding this new and unprecedented way of managing distances: "Going from maps of the world to the domestic appliance department in department stores (television aisle), writes Régis Debray, planet Earth was simultaneously miniaturised and domesticated. It can now be delivered to your doorstep, like a fridge or a vacuum cleaner)".[33] This is not an exaggeration when we think of the craze sparked by GoogleEarth and other sites offering us images of any location on Earth seen from space. However, are satellites not just *Big Brother* tools belonging to several political and economic powers who direct the world using the means made available by space technologies and other resources?[34]

5.1.1 Satellite voyeurism

Every square metre of the globe is now under top surveillance. These eyes belong not only to military satellites but also to civil remote observation satellites, all orbiting at an altitude of several 100 km, in other words out of reach and practically invisible. Of course, meteorological images help us plan our leisure and holiday time and GoogleEarth allows us to admire a multitude of views of the Earth from space. Yet how can we escape the feeling of being watched without our knowledge, of being spied on, not only by professionals but now by anyone who lives on Earth and has access to the Internet? Satellite voyeurism is not just myth: in fact, one of the first studies into the ethics of space activities focused on precisely this question: "*Aspectos éticos de los satellites*" (Ethical aspects of satellites, by Robert S. Hartman, published in 1959).

Circumterrestrial space is an environment that is particularly well suited to observation and intelligence, using sensors placed on board satellites that move around it. The first Earth observation satellites were launched as of 1960 for meteorological purposes. The United States launched its first spy satellite Corona in August 1960, and placed the first civil remote sensing satellite, Landsat 1, in orbit on 22 July 1972. Many countries now have their own Earth observation satellites.

The term "resolution" is often used when talking about surveillance satellites in general and in particular about spy satellites. This term designates the size of the smallest object that can be detected and recorded. Satellite resolution capacities have vastly improved over the last 50 years. The resolution capacity of the first instruments was several dozen metres whilst that of the KH-12 American spy satellites is reduced to 10 cm. Is there a theoretical limit? Specialists often speak about 2–5 cm due to atmospheric disturbances which cannot be corrected at such a low scale as resolution. These instruments are therefore installed on board satellites that are "fellow travellers" (meaning of Sputnik in Russian) of the Earth: they are all moving around our planet, following multiple trajectories. The choice of orbit is directly dependent on the satellite's appointed mission and the goals it must achieve: the altitude at which it must operate, the size of the image it must provide, the frequency of its crossings over the same territory, the areas that it must fly over, the brightness it requires, etc. Orbits can be classified in many different ways. They can be classified according to their eccentricity (they can be circular or elliptical, but also hyperbolic or parabolic for vehicles that escape the pull of the Earth), according to their synchronicity in relation to the Earth's rotation or according to their altitude. Based on this last criteria, four classes of orbit are currently distinguished: LEO (Low Earth Orbit), MEO (Medium Earth Orbit), GEO (Geostationary Earth Orbit) and HEO (High Earth Orbit). Low orbits and geostationary orbit are more directly associated with observation and surveillance missions. Located a few 100 km above the Earth, low Earth orbits are suitable for obtaining high-resolution and very high resolution space and radiometric data. However, compromises must be made. A very high resolution (of about a metre or even a few decimetres) is most often associated with a low scan swath and infrequent travel over the same point. On the other hand, an average resolution permits a higher scan swath and more frequent travel over the same area. Geostationary orbit, still referred to as Clarke orbit, is rather special and of particular use: satellites appear immobile at an altitude of 35,786 km. In fact, they travel at a speed greater than 10,000 km per hour in the equatorial plane and make, like Earth, a complete orbit in 23 h and 56 min. Used in particular to retransmit telecommunications, geostationary orbit is also conducive to the study and monitoring of rapidly varying phenomena, in particular meteorological phenomena.

And what about management of the data thus acquired? With such technical capacity, it is not difficult to accuse civil or military satellites of voyeurism. European farmers, whose sown areas, fallow areas and harvests are controlled using images from space, have in the past readily viewed Spot as an informer and remote sensing as a sophisticated form of remote sanctioning! Is that saying that the implementation of remote sensing systems and their use cannot be controlled or regulated in any way?

In fact, it must be recognised that there was no pre-existing legal structure before remote sensing was developed. For example, the Outer Space Treaty does not take into account the question of the public or private nature of these activities. When in 1973 the United States transferred the responsibility of their remote sensing system from NASA to the National Oceanic and Atmospheric Administration (NOAA), and thus brought remote sensing in closer relation to meteorology, the operation seemed to be pretty insignificant: meteorological images are above all intended for government agencies, while satellite images may concern a much wider and more economically varied group of users. That being the case, should these images remain free? Through the *Land Remote Sensing Commercialisation Act* of 17 July 1984, the development and exploitation of remote sensing satellites are entrusted to private companies, thus permitting the commercialisation of data. However, as long as the Cold War continued, no civil application of remote sensing products and technologies was permitted, whilst the resolution was less than 30 m. The launch of Spot 1 in 1986 marked the end of this limitation, this satellite offered commercial images with a resolution of 10 m from the onset. The following year, in 1987, the Soviet Union started to declassify and sell images offering a resolution of 5 m. In 1992, this resolution became 2 m. The residual American reluctance gave way through fear of seeing the satellite image market monopolised by other countries. The American authorities were forced to lift the various bans imposed on remote sensing activities and thus to allow national companies to play an active part in this new commercial sector, but not without stipulating restrictive clauses in relation to the most sensitive data or regions of the world. The *Land Remote Sensing Policy Act* authorising the sale, under licence, of satellite images was passed in 1992. This first step was followed on 23 March 1994, by the decision of President Bill Clinton to allow private companies to develop and place into orbit or sell very high resolution Earth observation systems. This resolution, currently limited to 41 cm, will be reduced to 30 from 2012. However, the American government reserves the right to prohibit the sale of images of sensitive areas (military areas, scenes of conflict). It was in the same vein that, in 1996, the United States Congress voted in favour of an amendment to the *Defense Authorisation Act*, enacted the previous year. This text prohibits the distribution of images of Israel taken by an American satellite with a resolution better than that of international competitors. Professionals ask the question: up to what point is it possible to implement and accept such a policy of *black out*? Who could prevent other countries from demanding the application of a similar restriction for their own countries? What then drives competition, when the United States imposes such a limit upon itself? In the meantime, on 22 February 1995, Bill Clinton authorised the declassification of data from several spy and intelligence programmes, including the famous Corona and Argon programmes. Over 860,000 declassified images were placed in national archives.

Spurred on by this lifting of the veil of secrecy, albeit very controlled, several American aerospace industry companies wasted no time in venturing into the world of very high resolution remote sensing. Late 1999, the company Space Imaging, established by Lockheed in 1994, started to sell images with a resolution of 80 cm, taken with the Ikonos satellite. For its part, the company Eyeglass, which later became Orbimage, firstly launched the Orbview satellites before establishing GeoEye in 2006 following its acquisition of Space Imaging. In September 2008, it put the GeoEye-1 satellite in orbit. With a resolution of 41 cm, this is the commercial satellite that offers the sharpest image to date. Many other American and Israeli satellites now populate Earth orbits including EarlyBird, QuickBird and Eros, etc.

5.1.2 From open skies to the charter "Space and major disasters"

Whilst the remote sensing of oceans does not give rise to any legal difficulty as these are areas that escape any state sovereignty, the remote sensing of continents on the other hand comes up against the principle of State sovereignty, as practised on their territory and in the air space above them. After the Paris Convention (1919), the Chicago Convention (1944) granted States "may prohibit or regulate the use of photographic apparatus" in aircraft flying over their territory. When it is a question of space technologies, can a State gather information about the natural riches and resources of another sovereign State without having obtained the latter's prior agreement? Is it not up to the "remote sensing" State to ask for the prior permission of the State whose territory is being observed? In fact, abandoning the demand for necessary prior authorisation, States, including developing States, rally to the "open skies principle", to the principle of the freedom to take remote sensing images, even if some of these activities are of a terrestrial nature. This principle is based on the more general principle of the 1967 Treaty which specifies the freedom of use of outer space. It applies it to the remote sensing satellite that operates in an area ungoverned by any state sovereignty. Moreover, the right of sovereignty of a State on its territory shall not be confused with ownership rights. In other words, for States there are no rights to images or privacy, as can be recognised for individuals. As for the practice of spying, which rapidly became commonplace for American and Soviet powers, it cannot really be governed by a regulation . . . .

Nevertheless, following the 1967 Treaty, the United Nations General Assembly adopted a resolution on 3 December 1986, which defined remote sensing as "the sensing of the Earth's surface from space . . . for the purpose of improving natural resources management, land use and the protection of the environment". This text poses a principle of distribution without discrimination and simultaneously establishes a form of responsibility, of obligation of States operating satellites or remote sensing systems, to help the territories and countries observed through cooperation, to share with them informations which have been gathered, processed and interpreted, to pre-arm them against the consequences of certain natural and manmade disasters, etc. With this resolution in mind, how exactly responsible are remote sensing States in practice? Is it possible to talk about deliberate omission, of non-assistance to countries in danger? Nothing to date can affirm this, although several space agencies have drawn up a charter than truly captures the spirit of this resolution.

In July 1999, during the UNISPACE III international space conference in Vienna, the French and European space agencies, CNES and ESA, decided to coordinate their satellite data delivery and acquisition capacities. In theory, it was not a question of wanting to tackle the growing number of American companies marketing satellite images, but a question of being able to offer this data free-of-charge to countries struck by major natural or manmade disasters. Thus the international charter *Space and Major Disasters* was born. The Canadian space agency ratified the charter on 20 October 2000 plus, since that date, space organisations and other organisations from a number of countries including India, China, the UK, the US, Japan, Algeria, Nigeria and Turkey. This surprising coalition goes beyond the usual political and economic splits. Space agencies and national or international space system operators are most likely to become members of the Charter and civil protection, rescue, defence and national security organisations in any of the member countries of the Charter automatically become authorised users.[35]

What do we mean by a natural or technological disaster? Any situation that leads to the mass loss of human lives, significant damage to resources on a large scale following a natural phenomenon (e.g. a cyclone, tornado, earthquake, tsunami, flood, forest fire and volcanic eruption) or a technological accident (in particular, causing chemical or radioactive contamination). The signatory agencies to this Charter therefore agree, during such events and only at the explicit request of those in charge of the affected area, to acquire and to provide satellite data that may provide help to the populations of the affected areas; it is not worth entering into the Charter initiation procedure. The Charter aims to ensure a permanent watch (an operator is on call twenty four hours a day), to provide images to the people and services who need them as quickly as possible, to schedule the taking of specific images and also to ensure the relevance and honesty of the request.

Since its application in February 2002, the Charter has been put into action more than two hundred fifty times (September 2010, including forty six times alone in 2007). This result often surprises the general public due both to the effective capacity of these space organisations to pool fragile and costly resources and to the number of "major disasters" striking human populations (mostly floods, followed by hurricanes and typhoons, volcanic activity and earthquakes, and more rarely hydrocarbon spills or transportation accidents). No continent is spared and many countries have triggered the Charter. The first call came from the Meuse and Moselle valleys in France on 4 February 2002, the second, on the same day, from the Democratic Republic of the Congo and the third, on 9 April, from Afghanistan following an earthquake.

To those who would be sorry to see our terrestrial cradle not only under surveillance by space powers but now delivered to our doorstep, we must now reply that the Earth has also become more than ever our own, a sort of immense spacecraft which we have boarded as a crew and on which we must remain for a long time to come. To do this, we must learn to live together and must implement rules and charters. Was this the new and powerful idea predicted by Hoyle in 1948? I am ready to believe it is. In any case, it is definitely the idea formulated a little later: the global village.

5.2 The planetary village

At the start of the sixties, Marshall McLuhan made himself the apostle of the global, planetary village; in *The Gutenberg Galaxy* (1962), he wrote: "The new electronic independence re-creates the world in the image of a global village". Fifty years later, it is impossible not to notice that the phenomenon described by McLuhan continues to accelerate and to grow, to such an extent that we could eventually suffer from too much information with our capacity to pay attention reaching saturation point. Sociologist Dominique Wolton may ask himself whether, having become a technical reality, the global village is effectively a social and cultural reality or whether "the interests of communication industries is not too often confused with the philosophical and socio-historic reality of users of these communication technologies".[36] Alain Pompidou, for his part, notes how "the notion of 'planetary village' is misleading. It is a village where we can discuss, trade and work but in which the identification of those conversing may become impossible."[37] Space technologies have had a decisive influence on these internationalisation or, better still, globalisation processes.

5.2.1 From internationalisation to globalisation

Internationalisation is not a new phenomenon: its origins which lie in the strong desire to discover and accumulate wealth, the desire to dominate and the fascination for risk and adventure, are intimately associated with the history and the nature of mankind, always looking to make its fortune, its good fortune. Driven by the progress of technologies and sciences, this process was however marked, from the end of the 19th century, by a clear intensification of exchanges and a true internationalisation of the economy. This phenomenon took a radically new turn during the Second World War. Without betraying its roots, its characteristics and its networks, it benefited from the large-scale emergence of industrial groups and the constitution of an integrated and powerful capital market. Integrated: that is definitely the key word for this new form of internationalisation which requires the use of a different term to describe it: that of globalisation. Marc Abélès emphasises how "the use of the concept 'global' would appear satisfactory in summing up the level of integration and interconnection which has now been reached and which can be seen through the empirical perception individuals have, beyond their patriotism and their cultural identities, of belonging to a global world".[38] Each of us can now move between several reference points, from the most local to the most global, all in the context of what could be described as a compression of space and time, immediacy and simultaneousness, which seems increasingly difficult to escape. This is a dynamic which Régis Debray picked up on and analysed in his *Histoire du regard en Occident* (History of the Gaze in the West): "The telephoto lens, for example, and photographic enlargements have altered our awareness of 'detail', and satellite images, our mental to-ings and fro-ings between all and some".[39] This dynamic is no longer restricted to photographic practice. The concerns of imminence, of the immediate and day-to-day events combine and are based on a more general understanding and concern for reality, with the feeling of belonging to the same planet. It was not until the 20th century and the invitation issued by René Dubos, to think globally and act locally, that there was any real and effective meaning.

It is not a question here of asking whether globalisation, the current, accelerated, intensified and even outdated form of internationalisation, is no more than a specific form of capitalism or whether it is, on the contrary, inseparably linked with the development of modernity, in the context of a market economy; nor of asking whether it is irreversible... I would leave these questions to economists and sociologists, retaining and highlighting here one characteristic of the so-called globalisation processes: the coexistence of apparently opposed dynamics, i.e. those of planetarisation and territorialisation.

Planetarisation, which globalisation most often brings to mind, is particularly evident in the domain of communications and thus in the domain of media and culture, sparking all the associated concerns over loss of identity. What does the future hold for cultures at this time of internationalisation? A number of observers pose the question. How many of these cultures can be considered to have entered the phases of erosion and disappearance? New information and communication technologies (NICTs) are intentionally presented as factors of a revolution from which a new society should emerge, but at the cost of how many cultures? Even so, it would not seem utopian to consider a territorial reaction in the sociocultural domain, similar to that which already exists in the economic domain. Although globalisation gives rise to the emergence of international and global organisations and structures, it simultaneously promotes a reorganisation of the economic area, in terms of territory and region. Europe is a good example of such reorganisation, as national rivalries have been substituted by continental complementarities. Not forgetting that natural boundaries continue to exist such as rivers, seas, deserts and mountain ranges.... Thus, territorialisation phenomena seem (to be able) to be associated with the evident and imposing planetarisation processes, preventing cultural levelling and promoting cultural cohabitation and co-evolution. Wolton continues: "If space is the preferred domain of communication, then territory is its limit".[40] He thus reminds us that a technological revolution would not necessarily cause wide-scale reorganisation of the global structure of societies: physical, cultural or social reality can counter or at least limit the effects of technology.

The political erosion that occurs with globalisation would however appear unquestionable. There are probably many exogenous factors associated with this downturn. However, it has to be admitted that politics has also weakened internally and has even abandoned several tasks, which up to now it had responsibility. Is this because of a timorous reaction with regard to a new form of barbarity that lies within societies themselves? "All known forms of politics – cities, States, Empires –, writes Patrick Viveret, had the characteristic of pacifying the interior by directing aggressiveness to the exterior, toward barbarians, strangers and infidels. On a planetary scale, no external enemy to mankind allows the problem of violence to be dealt with using traditional forms of exorcism. The only radical threat to mankind today, is itself. If barbarity exists, it is inside".[41] This is probably the first time that mankind has questioned itself about the status of barbarians, of these "surplus men": Any imperialist venture is likely to regenerate it and to have to manage it. That said, Viveret is right to underline how, in the context of a planetary village and increasing globalisation, it is more difficult or even impossible to channel violence generated by our societies by directing it towards external barbarians . . . unless we turn to the hypothetical Martians or deport new barbarians to other planets.

Thus, in more ways than one, space, with its skills and its technologies, its projects and its realisations, yet also its hopes and its fears, its symbols and its dreams, is directly concerned by the social, political, anthropological and ethical questions, sparked and maintained by globalisation.

5.2.2 Space, provider without frontiers

Of course, space is primarily a provider of particularly effective means for implementing these globalisation, planetarisation and territorialisation processes. If the many satellites and parabolic antennae allow the North-American culture to expand beyond its original territory and within societies which were until now preserved, it also makes it possible to maintain the cohesion or reconstitution of dispersed cultural communities, of these diasporas that now exist in their multitudes; to such an extent in fact that these antennae can become the target of supporters of the social and cultural integration of emigrants within their host societies. Moreover, we must not forget the emergence of new traditions, of new cultural movements, due to the meeting, mix and hybridisation of previously existing cultures, which was up to now difficult or even impossible.

Beyond technologies, space can also provide the concrete experience of cohabitation. Has it not itself experienced the shift from an era of territorial or national opposition (that of the race to the Moon) to that of international collaborations (of which the *Apollo-Soyouz* mission in July 1975 symbolically marked the start)? Today, getting teams and crews of diverse national and cultural origins to work or even live together, is the reality of every space venture. The setting up and running of the international space station is the example that immediately springs to mind. Without doubt, the transition from a group of a dozen people or even several hundred people to a planet that is home to several billion would appear exaggerated; the processes of meeting, exchange and absorption of cultures which are imagined and implemented are no less real and worthy of interest. I am convinced that space can constitute one of the laboratories of our future terrestrial societies.

By way of illustration, I would mention here the experience acquired in the distribution of orbits and radioelectric frequencies; the issue is in effect that of the allocation of a limited resource. The International Telecommunications

Union (ITU) was made responsible for this task. From a technological point of view, the ITU firstly allocates the frequencies according to the different uses (allocation), then between the different States for each of these uses (allotment); then, state organisations offer frequencies that have been allocated to them to their different users (assignment). Up to now, the ITU has fulfilled its mission by applying the "first come, first served" rule. This rule has the advantage of making the best use, at least in theory, of the frequencies and orbits that are technologically more valuable. But this method of allotment raises the question of disparity between States at different stages of development in terms of space activities. By the time the least advanced States have the means to access space communication technologies, only the least worthy or most difficult to use frequencies and orbital positions will remain. To ensure justice, equality and equity, should we not establish an allotment of these resources between States independently of their needs or their current technological capacities? This approach was applied in a non-telecommunications domain for the distribution of frequencies reserved for live television. During the world administrative radiocommunication conferences of 1977 and 1983, each State was allotted a position on geostationary orbit and five frequencies, regardless of its needs and size.

However, television broadcasting raises other difficulties of an ethical nature. The transmission of television images via satellites concerns freedom of expression, at the same time as the respect of individuals, cultures and national sovereignty. Admittedly, article XIX of the *Universal Declaration of Human Rights* proclaims "freedom of opinion and expression" and the right "to seek, to receive and impart information and ideas through any media and regardless of frontiers". Article X of the *European Convention on Human Rights*, 1950, affirms, for its part, that "everyone has the right to freedom of opinion and expression. This right shall include freedom to hold opinions and to receive and impart information and ideas without interference by public authority and regardless of frontiers". Finally, articles XIX and XX of the *International Covenant on Civil and Political Rights*, 1966, also refer to the principle of freedom of expression, information and opinion. How far can this principle be applied to States, even mandatorily? In other words, to what extent does a State have the right to transmit or to authorise the transmission of television broadcasts to another State, without having obtained the prior agreement of this latter state? This question effectively raises the issue of the responsibility of States, within the context of space law, i.e. under the regime of state responsibility with regard to the international community as far as space activities are concerned. In other words and out of respect for cultural identities, States should be able to carry out minimal surveillance of private legal organisations that make it possible to broadcast programmes using space technologies, without regard to borders.

Nobody can deny that this is a subject of much debate. Whilst liberal States generally favour the application of the principle of freedom to broadcast (so long as they can practise a *de facto* monopoly . . .), other States are concerned about its political consequences (this has traditionally been the case of Eastern countries) or cultural consequences (not only Third-World countries, but also other countries such as France). The latter group would be in favour of imposing an obligation on broadcasting States (the State of registration of the spacecraft used, the State which assigns the frequencies or the orbital position) to request the agreement of the receiver States. Be that as it may and generally speaking, including questions which concern the infringement of privacy, Mafia and terrorist organisations, violence or pornography, an international regulation is still lacking. For the time being, the ethical debate surrounding telecommunications of any nature essentially falls within the legal domain, focusing on the compromise to be found between the principle of liberty (both in terms of broadcast and reception), and the possibility of retransmitting events of major importance (and to avoid, for these specific cases, the exclusivity of subscription channels) on the one hand, and the protection of specific cultural issues (e.g. European productions competing against North American productions), the protection of minors and law and order, and the right to reply, on the other hand.

Far be it from me, of course, to suggest or claim that space would offer only benefits to modern societies taking advantage of globalisation; as they are also confronted by its restrictions and its misdemeanours. To comprehend the value of the charter on *Space and Major Disasters*, how many technological programmes exist that aim to provide a state or industrial organisation with even more power or to enable it to preserve this? The fact that the usual distinction between civil space and military space now seems to be blurring seems to me to offer another example of this development and these issues.

5.3 Grey space

In the beginning, everything seemed quite simple. Space was a product of the volition, the interest and the commitment of the military and scientists, of "the broadsword and the lab bench", as Roger Bonnet articulated so well in *Les horizons chimériques*. It all goes down to the work of engineers, who, since the end of the 19th century, gradually developed the main astronautical capacities and technologies, from those of launch vehicles to those which control the movement of a satellite or a probe in sidereal space. This role assigned to engineers found its image, or its incarnation, in Sergueï Korolev who was entrusted with managing and being responsible for the entire Soviet space programme. He hid behind the single and anonymous title of *Chief Designer* for a long time with the task of managing civilians and the military. Only in the Soviet Union was the merger (or should we say confusion?) of these two domains taken so far. On the contrary, elsewhere, the military tried to keep the control of some technologies to itself, whether these technologies were designed to develop weapons or surveillance systems. Space was then described as dual, in the etymological sense of the term: two opposing yet equally necessary principles. In other words, space applications came from either the civil or the military domain, with no possible or permissible confusion.

"Black programmes" or top secret projects which their governments and industrialists still keep under wraps for as long as possible, were and still are common in the military sector. Civil projects mostly fall within what are classified as "white programmes". The existence and content of the latter are readily made public, but there is an increasing number of programmes that can be qualified as "grey programmes", i.e. those that combine the two genres. Neither black nor white, space now seems to be increasingly overshadowed by the colour grey. In addition to other uses of space technologies, Earth observation and the monitoring of the activities of its inhabitants are also necessarily concerned by this change in colour.

5.3.1 Secrecy and democracy

Politicians are all too aware of this dilemma. It is not always easy to know exactly how much information about military programmes to divulge to citizens or, conversely, how much with regard to national intelligence and surveillance activities should be kept strictly confidential. In this respect, discretion, in the sense of due measure and reason, is a quality that requires sensitive handling. Without even mentioning democratic motivations, is it not necessary to encourage, or even to stimulate the interest and support of public opinion in favour of programmes that are a huge financial burden to a State? The latest generation American spy satellites cost at least one billion dollars each. However, the economic cost is not the only concern of citizens: how much blind support should they in effect give to intelligence activities which target not only potential or recognised enemies, but also themselves and their families for all they know? So long as spying and intelligence remained limited to wartime activities, as in the era of the famous Chinese general Sun Tzu or during the Middle Ages, secrecy barely posed a problem or, was at least part of a specific practice or ethic. This is no longer the case now that these activities are practiced in times of peace and extended to other areas of life: social, economic, political and scientific. Are excessive secrecy and the effects of subsequent misinformation compatible with the democratic perspective and project? These questions have become increasingly difficult to answer today due to the concomitant weakening of notions such as patriotism and the enemy and above all, due to the expansion of what I refer to above as "grey space". What becomes of secrets, democracy, the individual and private sphere, when civilians and the military start sharing more and more technology and data?

5.3.2 An over-exploited sky?

Space cannot escape these difficulties and these questions. To a history already rich in veiled sentiments, undercover programmes, deliberately blurred images, divulged secrets and condemned spies, is now added the enthusiasm of private companies claiming to offer to any of their customers images that General Eisenhower (to adopt the advertising slogan of one such company) would have loved to have got his hands on to prepare for the Normandy landings of June 1944. How is it possible not to ask questions or raise concerns about this development of technologies and in particular practices that no longer reserve information and intelligence for the military and government authorities, but disseminate it instead amongst an increasing number and diversity of users, who become more and more difficult to identify? By authorising the sale, at little or no cost, of very high resolution images which, just fifteen years ago were still classified and protected by defence secrecy, are governments playing into the hands of terrorists? Should we prepare ourselves for a fresh outbreak of armed conflicts and terrorist acts? Or, on the contrary, could the process of information democratisation, in this case the sale of satellite images, be a stabilising factor in the same way as the cross spying of the super powers resulted in a slowdown in the arms race and nuclear proliferation during the Cold war? At the end of the 1990s, journalist Charles Lane wrote: "Islamic Jihad could get its hands on a 1-m resolution picture of... a US Air Force General Headquarters in Turkey, convert the shot to a precise three-dimensional image, combine it with data from a GPS device... and transmit it to Baghdad, where a primitive cruise missile, purchased secretly from China could await its targeting coordinates[42]". Experts are pessimistic and, in addition to the dangers associated with terrorist activities, see the decrease of government leeway during negotiations conducted during international crises: does increasingly shared knowledge equate to diluted power and less authority?

5.3.3 Civilians and soldiers, fighting the same battle?

Many experts wonder whether it is possible to talk about a militarisation of space. Since the 1967 Treaty stated that the Moon and other celestial bodies could be used "exclusively for peaceful purposes" and officially prohibited the "testing of weapons of any kind and the conduct of military manoeuvres" yet contented itself with prohibiting the "placing in orbit around the Earth any objects carrying nuclear weapons or any other kinds of weapons of mass destruction", the question has been asked time and time again. In actual fact, the armed forces mainly use space for immaterial purposes, i.e. they use it to gather and transport information, from or via spacecraft: spying on the one hand, i.e. observation and listening; transmission on the other hand, i.e. telecommunications. It is worth looking at the scale of this use. Thus, as part of the Global Information Dominance programme, the United States set up the specialist National Imagery and Mapping Agency (NIMA) comprising 9000 people responsible for centralising and processing all images obtained by military satellites (in particular the Key Hole and Lacrosse series, whose resolution would reach 10 cm). For its part, the National Security Agency (NSA) employs 38,000 people of whom 20,000 are based in the United States. The aim of the National Reconnaissance Office (NRO) is as clear as its motto: "Freedom's sentinel in space. One team, revolutionising global reconnaissance". We are aware of the debate surrounding Britain's part in the Echelon spy network, as part of the UKUSA treaty that links the United States, the United Kingdom, Canada, Australia and New Zealand. This network relies on spying skills obtained through the alliance of the computer with the satellite (in particular the Trumpet and Vortex 2 satellites which allegedly have antennae 150 m in diameter). The Echelon network is said to be capable of spying, sorting, decrypting, archiving and processing three million telephone calls transmitted by satellite every minute, i.e. conversations, faxes, Internet messages and all electronic data exchanges.

However, we must note that the use of space equipment for military purposes remains regulated and restricted through international agreements, which prohibit the use of certain weapons, and due to political determination or technological constraints. The 1967 Treaty prohibits nuclear weapons and weapons of mass destruction in space; in Antarctica, this ban extends to all activities carried out in the name of national sovereignty and all military weapon testing. In any case, space serves only as a crossing point for intercontinental missiles without its peaceful status being legally violated. The ABM or Anti-Ballistic Missile Treaty prohibits the "development, testing, or deployment of sea-based, air-based or space-based ABM systems and their components, along with mobile land-based ABM systems".

The 1967 Outer Space Treaty and the issue of militarisation

"*The States Parties to this Treaty,*

Inspired by the great prospects opening up before mankind as a result of man's entry into outer space,

Recognising the common interest of all mankind in the progress of the exploration and use of outer space for peaceful purposes, [. . .]

Desiring to contribute to broad international cooperation in the scientific as well as the legal aspects of the exploration and use of outer space for peaceful purposes, [. . .]

Recalling resolution 1884 (XVIII), calling upon States to refrain from placing in orbit around the Earth any objects carrying nuclear weapons or any other kinds of weapons of mass destruction or from installing such weapons on celestial bodies, which was adopted unanimously by the United Nations General Assembly on 17 October 1963,

Taking account of United Nations General Assembly resolution 110 (II) of 3 November 1947, which condemned propaganda designed or likely to provoke or encourage any threat to the peace, breach of the peace or act of aggression, and considering that the aforementioned resolution is applicable to outer space, [. . .]

Have agreed on the following: [. . .]

Article III

States Parties to the Treaty shall carry on activities in the exploration and use of outer space, including the Moon and other celestial bodies, in accordance with international law, including the Charter of the United Nations, in the interest of maintaining international peace and security and promoting international cooperation and understanding.

Article IV

States Parties to the Treaty undertake not to place in orbit around the Earth any objects carrying nuclear weapons or any other kinds of weapons of mass destruction, install such weapons on celestial bodies, or station such weapons in outer space in any other manner.

(*Continued*)

The Moon and other celestial bodies shall be used by all States Parties to the Treaty exclusively for peaceful purposes. The establishment of military bases, installations and fortifications, the testing of any type of weapons and the conduct of military manoeuvres on celestial bodies shall be forbidden. The use of military personnel for scientific research or for any other peaceful purposes shall not be prohibited. The use of any equipment or facility necessary for peaceful exploration of the Moon and other celestial bodies shall also not be prohibited."

Without delving into the whys and wherefores of law and terminology, it is relevant here to distinguish between space weaponisation and space militarisation. In terms of deployment and implementation of conventional weapons, it is accurate to say that space is not the scene of true weaponisation. However, the United States appears willing to or is in the process of bringing this about, spurred on by their desire for space dominance and dominance using space means. However, we have to admit the process of space militarisation is already well underway, since this term designates the process which results in the direct contribution of space resources to force projection operations and to run military operations.

Space was essentially of strategic interest during the thirty years preceding the first Golf war. However, the situation changed due to the growing importance of information in the use of modern weapons. Military staff recognise that information management guarantees more than ever the freedom of action of armed forces. And so the conclusion is inevitable: the United States has clearly militarised space to its advantage. Other countries, including France, are currently following the same process, for planning and conducting operations. John Logsdon, an expert in space policies and affairs, is right to conclude: "It may well be that the time has come to accept the reality that the situation of the past half century; during which outer space has been seen not only as a global commons but also as a sanctuary free from armed conflict, is coming to an end". I think we should even go further and question the established and comfortable idea of the duality of space, in other words a strict separation between civil and military domains.

Is it still possible to speak of dual space, when the armed forces start regularly using the services of civil remote sensing systems? Iraqi forces during the conflict with Iran, American forces during operation Desert Storm, and NATO forces intervening in Bosnia and Serbia all used Spot satellite images. Is it still possible to talk about dual space, when GPS, originally intended for the American military, has become a space tool with innumerable civil applications? How does the concept of dual space fare when the future French observation system Pleiades will provide images to military intelligence, assuring the total confidentiality of their requests, secure communications, but above all absolute priority over requests from the civil sector? In short, once the barrier between civil and military is lifted, what is left of space duality? Perhaps "twinning" would be a more appropriate word, admitting the existence of similarities and complicity?

I believe that formal recognition of such a change should lead to the assignment of new tasks to civil space agencies. These new tasks include ensuring compliance with the choices listed and ratified in 1967, informing leaders and the general public of the consequences that some military practices may have on the peaceful nature of space use (e.g. the installation of lasers on board satellites), and monitoring the possible use of data gathered and sent by civil satellites to help terrorists or aid criminal practices (as evoked earlier). As we know all too well, this latter task is complicated by the fact that space appears here to be a metaphor for capitalism: the possession of the technology or the money can in itself bestow power, ungoverned by any legal, regulatory or political constraint. Ethical thought is still in its very early stages in this domain.

[31] Taken from Russell, Peter. *La Terre s'éveille: les sauts évolutifs de Gaïa.* Barret-le-Bas: Editions Le souffle d'or, 1989: 19.
[32] De Certeau, Michel. *L'invention du quotidien. I. Arts de faire.* Paris: Gallimard, 1980: 140.
[33] Debray, Régis. *Vie et mort de l'image. Une histoire du regard en Occident.* Paris: Gallimard, 1992: 412.
[34] See Arnould, Jacques. *La Terre d'un clic. Du bon usage des satellites.* Paris: Odile Jacob, 2010.
[35] For further details, visit the website devoted to the charter: www.diasaterscharter.org.
[36] Wolton, Dominique. *Internet et après? Une théorie critique des nouveaux médias.* Paris: Flammarion, 1999: 130.
[37] *Ibid.* 130.
[38] Abélès, Marc. *Anthropologies de la globalisation.* Paris: Editions Payot & Rivages, 2008: 8.
[39] Debray, Régis. *Vie et mort de l'image.* 179–180.
[40] Wolton, Dominique. *Internet et après?* 184.
[41] Viveret, Patrick. "Agir dans la mondialisation". *Esprit* 11 (1996): 130.
[42] Quoted in Steinberg, Gerald M. "Dual Use Aspects of Commercial High-Resolution Imaging Satellites". *Mideast Security and Policy Studies* no. 37, 1998. (http://www.biu.ac.il/SOC/besa/publications/37pub.html, January 2010)

CHAPTER 6

6 A threatening sky

Ensisheim, 7 November 1492, in the morning. Preceded by a terrible explosion, a black rock weighing approximately a 150 km flew across the skies over Alsace, in north east France, eventually landing in a field near the city walls. A young boy, the only person to have seen the rock hit the ground, guided the curious townsfolk to the point of impact: a crater two metres deep! Not really scared, the crowd eagerly helped themselves to fragments of this "rock fallen from the sky" for good luck. The pillage was luckily stopped by the bailiff, who ordered the rock to be taken to the church's doorstep. After all, it did come from the heavens. Nineteen days later, the young king Maximilian, heir of the Habsburg family and future emperor of Austria, came to the town of Ensisheim. He asked to see the rock and regarded it as a miracle and a good omen for the war he had to lead against the French. He ordered the rock to be kept in the church's chancel where it would remain for three centuries.

The French like to tell us that their ancestors, the Gauls, were intrepid warriors and only feared one thing: the sky falling on their heads. Were they really the only ones to be frightened by the spectacle of a meteorite shower? I doubt it. Be that as it may, it is important to give the Ensisheim episode a strong symbolic character. On that day, it became clear that the sky was not as stable, fixed and unchanging as many human conceptions, representations and cultures had believed. The sky, like the Earth and mankind, can be changed, damaged and broken. Now seemingly more accessible, the sky also posed more of a threat to humans.

6.1 The cluttered sky?

If you were to ask the so often solicited "man in the street" about the ethics of space, he would without doubt begin by admitting his ignorance or try to hide it behind the conspicuously affirmed opinion that space poses no ethical problem; unless he confesses that all these questions go way over his head, both figuratively and literally speaking. However, ask him about space debris and, if he is a regular enough reader of the daily papers, there is a strong chance he will change his discourse and show concerns about all of these objects, now redundant and unusable, hurtling around in the skies above us. Is it a trace, or a rekindling of the fear that the sky will one day fall on us? Why not. After all, not so long ago people were still wondering whether the continued launching of sputniks and satellites into space by the Soviets and Americans would upset time and the seasons; not to mention the fears that an eclipse with a reasonable amount of media coverage can still create.... On a more serious note, back in 1962, joining the protests of astronomers, Bernard Lovell described the US Air Force Westford programme as "ethically wrong". As mentioned above, 350 million copper needles were sent into space in October 1961 to be followed by more in 1963. Each one constituted a dipole antenna which together was intended to form a passive circumterrestrial reflector to improve military communications. But, asked Lovell, what would happen to all of these needles? Would they not constitute a danger or pollution?

6.1.1 More and more objects in the sky

On 4 October 1957, the Soviet Union sent the very first artificial satellite, Sputnik, from Earth into orbit using a Semyorka rocket. What would happen once orbit was reached and Sputnik was released? Whilst the first and second stages immediately fell back to Earth, the third stage and the fairing, no longer serving a purpose, remained in the same space trajectory as the satellite and became the first debris, the first space junk. Between them, they weighed more than 6500 kg, whilst the Sputnik weighted only 84 kg. The payload thus scarcely represented more than 1% of the mass injected into orbit! After 21 days of loyal service, Sputnik remained in orbit for a further 92 days, itself no longer serving any purpose. It then returned to the Earth's atmosphere where it combusted. The world thus discovered that it was impossible to enter space, to move around it and to work there without producing debris.

Since the launch of the Sputnik, many space vehicles have been sent into space including manned spacecraft and orbital stations, satellites placed in an Earth orbit and probes sent into outer space in order to explore other celestial bodies, launch vehicles of all types, etc. Approximately 20,000 metric tons of material has been placed in orbit, in circumterrestrial orbit alone, over the last 50 years and 4500 metric tons remain there. This number of tons corresponds to 4800 satellites in total placed around the Earth; 2400 of these are still present even though more than three quarters of these have ended their mission. In 2009, the service life of 21 geostationary satellites came to an end: 3 were purely and simply abandoned, 11 were correctly placed into an orbit ensuring they presented no further danger to other satellites or spacecraft (I will expand on this later) and 7 were incorrectly deorbited. It is worth comparing the figures with those obtained between 1997 and 2000: of the 58 geostationary satellites that had reached the end of their service life, 22 were purely and simply abandoned and the 36 others were transferred to different orbits; only 20 of these 36 were transferred correctly.

But what exactly is space debris? The COPUOS technical and scientific subcommittee defines space debris as: "all man-made objects, including their fragments and parts, whether their owners can be identified or not, whether they are in Earth orbit or re-entering the dense layers of the atmosphere, that are non-functional with no reasonable expectation of their being able to assume or resume their intended functions, or any other function for which they are or can be authorised". Now, in concrete terms, what exactly is this debris?

First and foremost it concerns operational satellites, around 600 of them, followed by those that are no longer in use. These currently represent 22% of the 14,000 objects catalogued, i.e. observable and monitored from the ground, in general of a size greater than 10 cm (we would doubtlessly need to add to this number a few thousand that are classified). The upper stages of launch vehicles that are used to place the satellites in orbit represent 13% of space debris. Technological and operational debris, i.e. another 13%, are objects voluntarily released during missions. These include covers used to protect instruments during the launch phase, systems used to fasten solar panels or antennae prior to their deployment in orbit, pyrotechnic fasteners and separation devices, straps, objects abandoned by astronauts during their extravehicular outings, etc. Finally, more than half of the objects catalogued are the result of the fragmentation of debris produced by the collision of an object in orbit with other objects and space debris or with meteorites, or which occur following the unintentional or deliberate explosions of space vehicles in orbit (e.g. military satellites that have failed to reach the planned orbit and which may accidentally fall into the hands of a foreign power).

Propellant residue must also be added to these objects. This includes solid propellant motor used for orbit transfers, in particular between a transfer orbit and geostationary orbit, releasing small alumina particles. Drops of sodium potassium leaked by old Soviet satellites whose electrical power supply came from a nuclear reactor also fall into this category. Last but not least, the ageing of material in space: the space environment is very harsh (sudden temperature changes between the shade and the Sun, atomic oxygen, ultraviolet rays, etc.) and therefore causes many alteration processes (separation of photoelectric cells, weathering of thermal protection covers, scaling of paints, etc.) which produce small debris.

The amount of debris is space regularly increases and since 1957, the number of additional catalogued objects per year is approximately 220. Between 2001 and 2006, a slowdown was observed. Optimists saw this as the direct positive result of prevention measures already applied by some operators, in particular, the passivation of satellites and launch vehicles at the end of their mission, in order to reduce the risks of later explosion and any subsequent fragmentation processes. Unfortunately, the hope of seeing this slowdown continue was shattered in 2007, with the explosions, intentional or accidental, of the Chinese satellite Fengyun 1C and the Russian stage Briz M.

On 11 January 2007, China tested an anti-satellite weapon in space. The old Fengyun 1C satellite was destroyed during an intentional collision with a missile launched from the ground. This created a cloud of debris that became distributed around the Earth. The American military space surveillance network catalogued more than 1500 objects larger than 10 cm in size and there are clearly many more pieces of debris smaller than this. This event occurred at an altitude of 850 km which is already one of the most polluted regions of space. Most of this debris will remain in orbit for more than two centuries. The other explosion occurred on 19 February 2007, when the upper stage Briz M of a Russian launch vehicle accidentally exploded producing over 1000 catalogued pieces of debris. This launch vehicle was supposed to place the Arabsat 4A satellite in orbit on 28 February 2006. Following a fault on the launch vehicle, the satellite and the upper stage remained in an intermediate orbit of which the perigee was at an altitude of 500 km and the apogee at 15,000 km. This explosion was caused by a mix of residual propellants and was therefore the result of a failure to passivate the launch vehicle at the end of its mission. The debris created on this occasion regularly crosses the most cluttered areas of low earth orbit and thus increase the risk of collision with operational satellites. It is interesting to note that, following the incorrect injection of the satellite, significant efforts were made to deorbit Arabsat 4A but to no avail. The efforts made to remove this single object from space must be considered with the thousands of pieces of debris produced by pure negligence in mind.

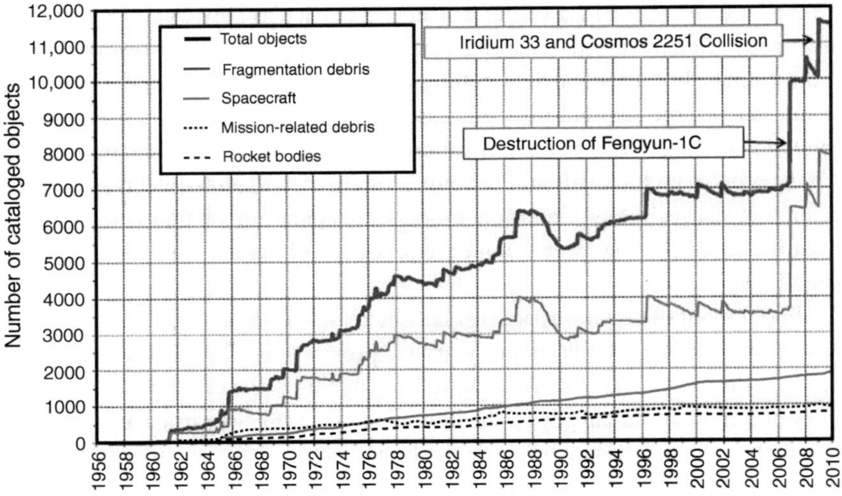

Fig. 1. *Number of catalogued objects around the Earth (source: NASA).*

Future development depends essentially on the number of launches, in particular on their decrease observed in Russia in recent years. It is also important to take into account the change in types of mission. Constellation projects involving the use of dozens, even hundreds of satellites could raise concerns over a rapid increase in the number of objects in circumterrestrial orbit. Most of these projects are currently dormant. Another potential risk is emerging with the development of "small" satellites (minisatellites weighing between 100 and 500 kg, microsatellites between 10 and 100 kg, nanosatellites between 1 and 10 kg and picrosatellites under 1 kg) which can be launched in clusters or swarms from the same launch vehicle. Of course, a final factor is the effectiveness of prevention measures.

6.1.2 Risk assessment

On 24 July 1996, the French electronic surveillance microsatellite Cerise struck a fragment from the third stage of an Ariane rocket launched ten years previously. The satellite gravity gradient boom (a boom approximately 5 m in length designed to passively stabilise the satellite) was fractured rendering Cerise inoperable. It was the American military, the only people at the time to have the adequate surveillance means, who informed the French authorities. Was this the first collision in space? Probably not. However, the Americans would doubtlessly have been more reluctant to inform the rest of the world about any less politically correct collisions, between American, Soviet or Chinese spacecraft.

On 10 February 2009, two satellites collided at an altitude of 790 km: an American satellite belonging to the Iridium constellation and an inactive Russian

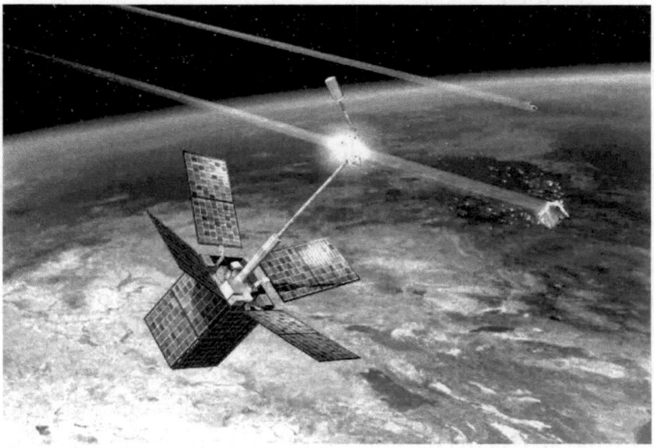

Fig. 2. *The Cerise satellite accident (source: CNES).*

satellite from the immense Cosmos series. Two clouds of thousands of pieces of debris quickly formed. Their apogees and perigees are now between an altitude of 250 and 1350 km. This translates into increased risk of further collisions for spacecraft and humans orbiting the Earth at these altitudes.

These two accidents are answer enough to those who ask themselves whether it is really necessary to be worried about space debris orbiting the Earth. The first and main risk associated with space debris is that of the collision with other objects in orbit which are not necessarily debris. If space debris moved around the Earth in an ordered way, it would not pose any real threat to human space activities, except for literally covering the space horizon .... However, this is not at all the case: these objects move in random directions like "bees disturbed from their hive" and can therefore collide with each other or with manned or operational spacecraft. Proportional to the size of the debris, the seriousness of a collision depends on the relative velocity of the debris in relation to the structure it hits. The travel speeds in question are considerable: in low earth orbit, satellites and debris move at a rate of 7–8 km/sec, i.e. 25–30,000 km/h. Relative velocities can thus reach and even exceed 10 or 15 km/sec. Such speeds involve vast amounts of kinetic energy and structures can easily be punctured or pulverised on impact: an aluminium sphere with a diameter of 1.3 mm, launched at 10 km/sec, has the same energy as a bullet from a 22 long rifle. In other words, a satellite may be destroyed by debris with a diameter of one centimetre and badly damaged by a grain the size of 1 mm.

In June 1999, the core of the International Space Station, then formed by the first two Russian and American elements, Zarya and Unity, narrowly escaped disaster, when a piece of debris, picked up late by surveillance telescopes and radars, passed dangerously close. There was no time to instigate an avoidance procedure. Since this date, the ISS has performed several course alterations for similar reasons. In fact, lots of debris floats around an orbital station yet poses no threat due to their being in the same orbit and the low relative velocities. However, when disturbed and according to the laws of celestial mechanics, orbit planes can develop differently and become increasingly distant from each other. Two objects initially placed in neighbouring orbits can end up meeting on secant orbits, or on the same orbit but the opposite way round. Happily, the chances of a structure meeting debris on a secant orbit is very slim compared with the chances of meeting an item of debris in its own orbit: despite the amount of debris, the probability of collision thus remains low.

In February 1997, the crew of the American shuttle Discovery, during the second maintenance mission of the Hubble space telescope, was forced to alter its trajectory to avoid all risk of collision with another rocket fragment which was less than 1.5 km away from it. In 1991, the same shuttle had been forced to take evasive action to avoid the third stage of a Russian rocket. A similar operation occurred in 1992. On 12 January 1996, the Endeavour shuttle was forced to leave its trajectory to avoid an American military satellite not in service. NASA directives specify a distance of at least one mile (1609 m) from a detected object. However, even a flake of paint can be dangerous for a space shuttle. Like a car windscreen, their main windows act as excellent impact detectors. Small impacts, e.g. those provoked by paint debris, do not constitute any real danger during atmospheric re-entry, at the end of a mission. However, during the next flight, due to the enormous stresses to which the spacecraft is subjected, these miniscule impacts can constitute fragile points and have detrimental effects on crew safety. It is for this reason that on average one, sometimes two, of the eight main shuttle windows need to be replaced at the end of each mission. More generally, the risk of serious collision between a shuttle and an item of debris is estimated at 1 in 200. Debris is responsible for or at least involved in eleven of the twenty disasters scenarios which space shuttles can face.

In addition to the size and relative velocity of debris, another important factor to consider is how dense the collection of the debris is. Density varies considerably depending on altitude.

The density of the collection of debris appears at its greatest at around 850, 1000 or 1500 km, with, on average, one object per 100 million km^3. Above 1500 km, this density decreases the higher the altitude, except when approaching the altitudes of semi-synchronous orbits (20,000 km) and geosynchronous orbits (36,000 km), where debris is locally more crowded together. Other orbits, today

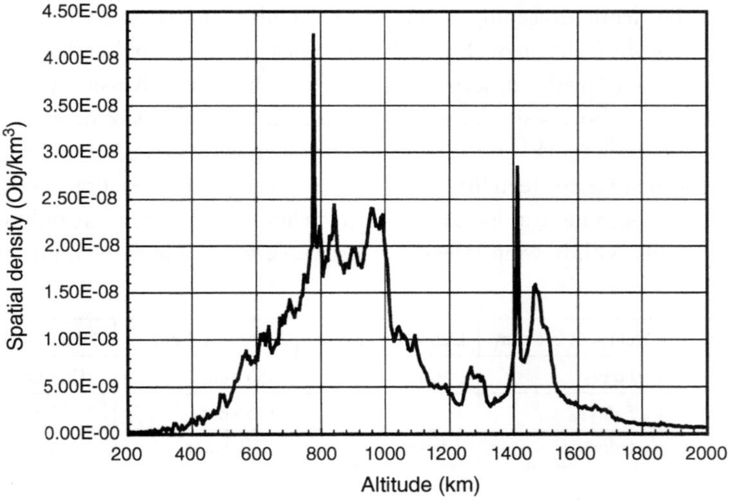

Fig. 3. *Density of space debris according to altitude (source: NASA) (translate ×0E-08 by ×.10-8).*

considered to be of no use and relatively free of debris, could one day change in status. Thus, orbits located around an altitude of 2500 km seem to interest the military in terms of observation satellite positioning. Wouldn't we be better trying to protect these orbits? Once the process of pollution begins, it becomes difficult to control it and impossible to reverse it completely.

Space agencies now have several ways, practices and theories, to study damage caused by debris and the frequency of collisions. The best way to observe "small" debris is to continue to use vehicles or elements that return to Earth after a stretch of time in space. In addition, missions have been specially dedicated to the study of debris. The LDEF (Long Duration Exposure Facility) satellite or the EuReCa (European Retrievable Carrier) satellite, which remained in orbit respectively between April 1984 and January 1990 and between July 1992 and May 1993, enabled a better understanding of the amount of debris and the number of meteorites. After replacement, the solar panels from the Hubble telescope were brought back to Earth and also provided a source of information.

These various studies led to a better understanding of the flow of particles, at least concerning specific orbits, and their probable origin (natural for meteorites, artificial for debris). The European space agency created the Master model which calculates the flow of debris around the Earth, based on the orbit and the size of the particles concerned. This data is used to assess the theoretical average times between two debris impacts.

Using similar processes, it has been estimated that the amount of debris in space would increase by 5% each year. In other words, at the end of the 21st century, there would be a 40% probability of a satellite in geostationary orbit being hit during its mission by debris exceeding 1 mm in size. Unless the rate at which debris is produced by the collision of objects already present becomes higher than the clean-up rate provoked by the atmosphere (see below), due to the "Kessler syndrome" in other words, a process similar to a chain reaction: two pieces of debris collide to produce new debris. When the concentration of objects exceeds a critical density (which is the case between 900 and 1000 km altitude), the reaction diverges and the increase becomes exponential. Experts believe that possibly the only solution would involve withdrawing at least five large objects per year from low earth orbit.

Altitude	Debris of 0.1 mm	Debris of 1 mm	Debris of 1 cm	Debris of 10 cm
400 km	10 days	3 years	885 years	12,900 years
780 km	1.5 days	1 year	155 years	1190 years
1500 km	1.6 days	1.6 years	270 years	1590 years

Fig. 4. *Average time between two debris impacts in LEO on a satellite with a surface area of 100 m^2 (source: ESA Space debris mitigation handbook, April 1999).*

But should we be alarmist? For example, at any given time there are only ten objects of ten centimetres or more simultaneously floating over France. In addition, by using another French marker, the total mass in orbit represents 1/4–1/3 of the mass of the Eiffel Tower, i.e. an average density of approximately 1 kg per million km^3. By way of comparison, the atmospheric density at an altitude of 300 km is 30 tonnes per million km^3. The probability of the ISS (which is the size of a football field) colliding with debris that is 1 cm or more in size is one impact every 70 years, of which only one every three centuries elsewhere than the solar panels.

Space Debris, the video game

"Fancy a game of Space Debris?" The name of the video game released on the French market in January 2000 could have pricked up the ears of those aware of the space debris dossier. However, in this case, they would have been disappointed. This game is simply about a series of pursuits on board a space vessel. The pilot (i.e. the player) must shoot at everything that moves to avoid his demise, and collect bonus points to increase the power of his shots and speed of his manoeuvres. So, why this name? Because of the debris moving randomly about, threatening to crash into the cockpit. Comment: "Keep your eyes open and stay alert like the dogs waiting to be thrown brown and orange snacks in television adverts". It's nothing to make a fuss about or to cause a wave of panic among the general public!

6.1.3 Monitoring debris

Collecting data from time to time or conducting statistical analyses on space debris is not enough. Space debris requires careful monitoring. In fact, along with nuclear waste, space debris is probably one of the most monitored sources of pollution. One of the main players in this surveillance is the American Space Surveillance Network (SSN) which owns several types of observation facility, distributed around the world, inherited from spy-satellite and missile detection systems installed by the United States during the Cold War. Radars are particularly suited to the surveillance of low earth orbits where they monitor all debris larger than 10 cm in size, whilst telescopes can be used to observe objects in geostationary orbit ranging from 30 to 50 cm in size. Optical observation can however be restricted by weather and lighting conditions. The object needs to be lit up by the Sun whilst the observation station remains in the shade.

Close to 150,000 observations are made daily and are used to update a catalogue listing approximately 9500 objects of varying sizes ranging from that of a baseball ball, to employ an American reference this time, to that of a Greyhound bus.

Europe has no specific space surveillance system although several space agencies and military organisations have developed and are currently developing radar and optical equipment. The French radar Graves (Grand Réseau Adapté à la VEille Spatiale – Large Network Adapted to Space Surveillance), developed by the French Aerospace Lab ONERA, is a radar prototype used to survey low earth orbits and to identify objects with a surface area equivalent to 1 m^2 at an altitude of 1000 km. Currently operated by the French Airforce, Graves is used to establish a catalogue of objects in low earth orbit, independently of the American SSN. Using its 34 m-diameter antenna, the German radar Tira, installed near Bonn, is capable of detecting and analysing 2 cm objects at an altitude of 1000 km. Combined with the Max Planck Institute's radiotelescope, it can reduce this size to 9 mm. In 2000, during a twenty four hour observation campaign, Tira detected 471 objects in low earth orbit of which only 94 were already catalogued. These two radars, Graves and Tira not only enable the orbit of a piece of debris to be determined, but also make it possible to validate environmental debris models for improving our knowledge of their populations. Europe also has other radars, other orbital debris/particle measuring instruments. All of these resources are used for making avoidance manoeuvres and for protecting satellites, even if they are costly for operators in terms of propellants and in terms of the availability of satellite resources.

A close call for the Spot 2 satellite

Just before 4 pm on 23 July 1997, the engineers in charge of Spot 2 discovered that there was a risk of collision between their Earth observation satellite and object 5061 (debris from a Thor Agena rocket), calculating that the collision could occur the following morning at 2:17am. Despite the unavailability of the French navy ship, Le Monge, whose radars are capable of tracking some space objects, the engineers decided to refine the calculation of trajectories; this only confirmed their fears. They consequently took the decision to perform an avoidance manoeuvre. That same evening, three orbits before the predicted collision, they ordered from Earth two 0.1 m/sec impulses in order to ensure a safety distance of 10 km. The next day, they reversed the manoeuvre to bring Spot 2 back to its initial orbit. Operation costs: two days without images and 400 g of propellants consumed (compared with 150 g for all position maintaining operations over the year).

6.1.4 Protecting orbits

Surveying debris that moving around above us or performing avoidance manoeuvres when this debris moves too close to our satellites, shuttles or station is not enough. If space agencies want to preserve future access to useful geostationary orbits, they absolutely must protect the orbits themselves against the proliferation, the exponential increase, of debris populations, and to extend the concept of sustainable development applied on Earth, to space. One of the principle measures to be implemented involves passivating launch vehicles and satellites once their respective missions have been completed and then, if possible, moving them outside of useful zones or zones designated as protected. These intentional and costly operations which, as we have seen in relation to geostationary satellites no longer in use, are not systematically performed by the agencies and companies that manage these space vehicles.

6.1.5 Protecting structures in orbit

Avoidance operations have already had to be performed and are regularly practised for orbital stations, shuttles and even some satellites. However, they can only be carried out for debris of a size that can be detected and monitored. What about that measuring less than 10 cm? The smallest of such debris, measuring less than 1 or 2 cm, can be stopped by shields with which space vessels can be equipped,

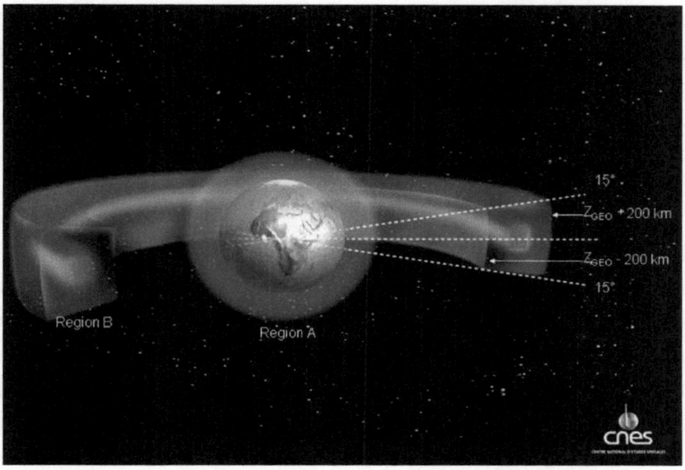

Fig. 5. *Protected zones in circumterrestrial space (source: CNES). Region A: low earth orbit (altitudes below 2000 km). Region B: geostationary orbit around which a corridor is added for station changes.*

especially those with multilayer structures. However, the most dangerous debris measures between 1 and 10 cm and debris of such a size is difficult to spot and monitor (radars use wavelengths greater than 10 cm and telescopes are not sensitive enough) and no shield can provide sufficient protection. There is estimated to be 200,000 objects of this size.

Also in the interests of protection, American shuttles fly, when their mission allows it, "head down" and "in reverse" to ensure that their strongest sections, i.e. their engines, meet any debris first and to protect particularly fragile sections such as the viewports and wing leading edges.

What about astronauts? In effect, when they leave the confines of their space vehicle, they are directly exposed to the risk of being struck by debris. Their spacesuits offer very little protection. However, their small surface area and limited duration of their extravehicular outings notably lessen the risk of collision. In addition, spacesuits are designed to ensure the safety of an astronaut for thirty minutes following the appearance of a hole that does not exceed one millimetre in size, i.e. the size of a pinhead. The greatest risks faced by astronauts are thus associated with spacesuit failure.

6.1.6 Reducing the debris population

Survival and protection therefore seem necessary, but is this enough? According to a strategy now implemented for most of our terrestrial activities, would it not also be advisable to reduce the number or production of this debris which constitutes real dangers for circumterrestrial space activities? In fact, space has its own self-cleaning system: due to the effect of friction with atmospheric particles, most of the objects launched into Earth orbit by humans end up by returning to Earth eventually. Travelling at very high speeds, the objects that do return to the Earth's atmosphere heat up and shatter. Most of the time, they are sublimated and leave only a trail of light in the sky. Some material, such as steel, titanium and ceramics, have a better resistance to heating and may reach the ground or, more often, the ocean intact.

As a first approximation, atmospheric density decreases exponentially with altitude. The atmosphere has no specific limit and, even in residual form, it can exert a braking force on vehicles closest to Earth and modify their trajectory. Subject to such conditions, the life span of an artificial satellite orbiting the Earth depends on the initial altitude at which it was placed. At an altitude of 350 km and in the absence of any heightening operation, the satellite is estimated to remain in orbital station for more or less one year. Spot satellites placed at an altitude of 825 km, will remain in orbit for 200 years. We are talking about several million years for objects in geostationary orbit. Thus, of the 25,000 space objects catalogued by the US Air Force Space Command between 1957 and 1998, 17,000 have now returned to the Earth's atmosphere. Each month, NASA publishes fallout forecast for the next 60 days.

There have been some large objects whose memorable fall out of orbit was controlled to a greater or lesser degree. These include that of the American Skylab laboratory, a 72-tonne craft that returned to Earth in July 1979. The most fragile parts (solar panels and antennae) were ripped off by the atmosphere and combusted over the Indian Ocean. Certain larger parts landed in Australia. The amount of debris that did not burn in the atmosphere and that fell out over a trajectory of several thousands of kilometres is estimated at 20 tonnes. The Soviet station Saliout 7 fell out of orbit on 7 February 1991 and some of its parts were found a few 100 km from Buenos Aires. The dates of these two spectacular fall outs are not totally accidental. Every eleven years, the Earth's atmosphere is subjected to a solar maximum, a period of increased solar activity. Subjected to increased solar energy, our atmosphere dilates, thus increasing its aerodynamic resistance and, consequently, the number of pieces of debris that fall back to Earth. In 1989 and 1990, during the last solar maximum, an average of three objects fell back to Earth each day compared with one per day in periods outside such peaks of solar activity. A total of 560 tonnes of debris was eliminated at this time, including the Saliout 7 station. The previous solar maximum had occurred in 1979 and 1980, causing the premature fall of Skylab. Independently of these effects of solar activity, the fallout of the 140-tonne Mir station received wide coverage. The Russians estimate that 10–20% of the initial mass ended up on the Earth's surface stretching over a trajectory of several thousand kilometres. The deorbiting was controlled to terminate in the South Pacific Ocean, bearing in mind that dispersion may occur, due to failures or atmospheric disturbances.

Atmospheric re-entry should be considered in terms of controlled or uncontrolled re-entry. During uncontrolled entry, the object (or what remains of it) falls anywhere within the band of latitude corresponding to the slope of its orbit. If the risk to populations is too high, controlled re-entry is necessary. Space agencies generally consider the acceptable risk threshold (i.e. the probability of claiming a victim during the operation) to be around 10^{-4} or a few 10^{-4}. One or more manoeuvres are required to make a station fall back to a zone where its fall runs a minimal risk. The Oceans that cover 70% of our planet are of course the "preferred" areas. In the case of the Mir station, the risk of claiming victims was estimated to be above the acceptability threshold and experts opted for a controlled re-entry.

However, deorbiting operations, i.e. a forced return to Earth, are costly. Controlled deorbiting at the end of the service life of a Spot family satellite, located at 825 km, requires approximately 150 kg of propellant; this is prohibitive.

The solution opted for by CNES for Spot 1 was therefore a return to Earth in under 25 years. In this case, the amount of fuel consumed is no more than approximately 54 kg, which is still a significant amount. Other potential difficulties include the complexity of modifications to be made to space vehicle design and a lack of propulsion resources. It is also important to keep in mind the negative, even disastrous consequences of adhering to such constraints in terms of commercial competitiveness.

The same type of constraints and criticisms impact other solutions sometimes evoked. Of course, some platforms and some satellites can be retrieved and brought back to Earth but this only occurs exceptionally. In practice, there is no solution that can be generalised to clean orbits by eliminating debris. Due to the great speed at which objects in orbit move, all capture systems (such as the "butterfly net", absorbent foam, etc.) would result, in the event of collision, in the disintegration of the interceptor and create more debris. With regards to larger debris, a rendez-vous is no easy matter when it concerns uncooperative objects at an uncontrolled attitude, probably rotating, nor is trapping and securing a potentially hazardous object (due to residual propellants, etc.) in a shuttle hold, and then having to perform an atmospheric re-entry with such a load. In addition, after a first rendez-vous, it is out of the question, due to the energy required, to modify the planned trajectory to go and look for another object located on another orbit. The cost of retrieving a single item of debris of any size in such a way is exorbitant.

"Star Wars" type destruction systems with high power lasers set up on the ground have also been considered. Above and beyond the fact that systems of this type are far from being feasible (in terms of the power required, pointing precision, uncertainty of the trajectory of debris), experts agree that it is preferable to have an object in orbit that is intact and that may be seen from the ground, rather than hundreds or thousands of potentially hazardous pieces of debris.

6.1.7 Space debris management with a long way to come

In itself, the existence of debris in space should not be surprising as all human activity inevitably produces waste in a variety of forms, whether predictably or unpredictably. Breathing, walking, eating, sleeping and all other acts and gestures that are a fundamental part of everyday life produce residues which, most of the time, we fail to notice or to which we are indifferent, at least while we are in good health or have enough living space and resources. The production of residues is simply a result of being a biological organism. However, if an anthill can manage the waste produced by its activity, our human societies should be able to manage waste even more effectively and more responsibly, to ensure our well-being and sometimes even our survival. Human activities are not solely derived from mechanisms inherited from past generations or imposed by nature, but inevitably comprise a dimension arising from free will, decision making and even the imagination. Cultivating a field is one thing, managing excessive agricultural production is another; cycling, taking public transport or driving a car do not have the same impacts in terms of waste management; building a hut, an administrative building or an Olympic stadium all have different implications, etc. So, anticipating pollution (even naturally produced), identifying it, assessing it, limiting it and isolating it are all part of the way in which mankind considers itself to be intelligent (does mankind not present itself as *Homo sapiens sapiens*?), free and a member of a group, a society, a species.

Space is not exception: space activity is a source of pollution, at first sight like any other, and space waste is a reminder of this. For a considerable amount of time, space waste was largely ignored or considered to be unimportant in relation to the space venture and its issues. However, it is vital that it be taken seriously from now on. That said, does it not in the first place lead to the same questions as all other forms of human pollution: who are the polluters and the victims? What is the scale of this pollution, in space and in time? What solutions can be adopted to solve the issue? Are these feasible? How much do they cost? Is it appropriate to apply the Polluter Pays Principle? What would the regulatory framework be, and how can it be enforced?

It is with this in mind that the main players in the space industry, including the heads of the national space agencies, put the issue of the technological and socioeconomic consequences of the proliferation of space debris on the table and attempted to implement a number of control, management and even ethical systems. It is mankind's moral duty, plus the necessity to protect astronauts and satellites, and to ensure the continuation of space activities around the Earth that forces us to deal with the issue of space debris, even if, strictly speaking, there is currently no law pertaining to the non-proliferation of orbital debris. We currently rely on the good will and the conscientiousness of users.

As one of the producers of debris and equally as a possible victim of the same (the Cerise and Iridium cases serve as a good example), CNES set up a "Space debris review group" in 1993, to define its short, medium and long-term policy regarding space debris and to inform all partners involved in the matter and in particular by regulatory aspects. Dealing with the question of space debris requires input from most space domain professions. At an agency like CNES, this means the involvement of several different areas of competence and departments: basic technologies such as space mechanics, measurement facilities, satellite technologies, quality or mathematical modelling, satellite operations, collision risk forecasting or the monitoring of atmosphere re-entries, satellite projects, launch vehicle design and launch operation strategy, relations with other space agencies and with the United Nations, legal issues; the Guiana space centre is also concerned by problems of collision risks at launch. One of the main products of this group was specifically the drafting and publishing of the document *Introduction aux exigences de sécurité relatives aux débris spatiaux* (Introduction to safety requirements pertaining to space debris) in 1998–1999. The foreword in this document emphasises that imposing good practice must not overshadow the constraints of the market and competition: "Considering the importance of the multiple consequences that its immediate application would bring about, consequences that would penalise CNES and its partners, it must be considered as a recommendation whose application remains the discretion of each programme or project. The application of a document of this kind will become mandatory once an international consensus has been reached [. . .] It will need to be referenced as a Standards document in all contracts between CNES and its partners. The requirements of this document will be requested at all levels of the organisation of a programme or project: partners, contractors, subcontractors, customers, suppliers, etc." This document was used to establish the CNES Standards Reference signed in 2001.

At European level, the CNES document was used to help draft a European standard, then transformed into a code of good practice, legally and ethically less restrictive. European agencies therefore now have a common reference document, defining the rules to be applied to satellites and launch vehicles in order to tackle debris proliferation.

Finally, at international level, the InterAgency Space Debris Coordination Committee (IADC) is made up of eleven members: ASI (Agenzia Spaziale Italiana – Italian Space Agency), BNSC (British National Space Centre), CNES, CNSA (China National Space Administration), DLR (German Aerospace Centre), the European Space Agency, ISRO (Indian Space Research Organisation), JAXA (Japan Aerospace Exploration Agency), NASA, NSAU (National Space Agency of Ukraine) and ROSCOSMOS (Russian Federal Space Agency). The Canadian Space Agency wishes to sit in as an observer on the next meetings. One of the key roles of IADC is to identify future preventive measures to restrict debris proliferation. With this objective, the IADC drafted its *Mitigation Guidelines* which summarise the opinions of its members concerning the measures to be applied. IADC has established itself as a driving force for the United Nations, the only establishment which would appear able to promote an international regulation which is still lacking today, due to the efforts of COPUOS and its legal and scientific and technical subcommittees. Its legal subcommittee has been dealing with the subject since 1996 and the second has not yet been asked for its opinion on the matter.

At the highest level, the United Nations decrees the principles that may flow down to national legislation or to the regulations of each country (terms and conditions for the issuance of licences for instance). This regulation is based on standards that allow manufacturers and operators to apply a common set of requirements that can be measured and checked. IADC, with its members actively participating at all levels of development, ensure the overall coherence.

With the support of IADC, the COPUOS scientific and technical subcommittee drew up a high-level document entitled *UN-COPUOS Mitigation Guidelines* which defines the main principles for measures to be adopted in space in order to reduce debris proliferation. It specifies a total of seven principles, which represent the consensus of the United Nations and which much serve as a basis for all national space debris regulations. This document was adopted by COPUOS in June 2007, and ratified by the United Nations General Assembly in December 2007 in its 62/217 resolution "International Cooperation in the peaceful uses of outer space".

In return, countries will need to implement such recommendations. The United States and the United Kingdom have already implemented a space activity licensing system which incorporates criteria pertaining to space debris. In June 2008, the French Parliament voted in favour of a law governing space operations (see before): the technical regulations associated with this law contain the requirements to follow, notably regarding space debris. Its content is directly inspired by ISO standard 24113 which merits international-scale recognition by space agencies and manufacturers and operators. These legislative mechanisms thus allow States to control the activities of their nationals by ensuring the compliance of space operations.

Looking at the facts, there is an undeniably clear and real need to define and improve the conditions of application of the first rules established to improve space activity management. We know that measures to transfer geostationary satellites to a graveyard orbit at the end of their service life are only being partially applied. Over the last few years, only one third of operators correctly performed these transfers, in accordance with IADC recommendations. Another third of operators partially completed operations with an insufficient reorbit altitude. The final third of operators did nothing, simply abandoning their satellites in geostationary orbit at the end of their service life. As we stated above, although space activity is increasingly in the hands of private companies, States clearly remain responsible for such activity and as such need to control space activities conducted from a national territory or by their nationals.

Industrial players (i.e. satellite manufacturers or operators) are not saying much. They claim to be ready to apply the recommendations drawn up by the agencies, insofar as, and predictably enough, all global players adhere to these rules. If they do not, competition between industrial players may be distorted. They also emphasise the need for agencies to support R&D programmes to develop reliable and effective deorbiting or reorbiting systems. How can the risks associated with atmospheric re-entry be managed? What type of design should be adopted for satellites? How can operations at the end of the life of satellites be controlled and decided upon?

From an insurer's point of view, it is important to remember that manufacturers can resort to them first and foremost to cover any damage that may be directly related to satellites, launch vehicles or launch operations. Up to now, no disaster has been caused by debris (the Cerise satellite was not insured). Space debris is not therefore their primary concern. Even so, insurers are aware of the current limits of their practices. The current terms of their policies ("legal liability" type) are 12 months, meaning that nothing is in place in the event of an accident after one year caused by debris or suffered by a structure in orbit. Even so, it would seem difficult for insurers to offer products with no time limit and no liability limitation. It is for this reason that the idea of a fund to cover damage caused by debris has already been evoked.

Legal expert from Brest (France), Armel Kerrest, likes to say: "We must not protract taking preventive measures in the hope that future technologies will allow us, like Hercules, to clean up the Augean stables of space". Taking seriously and taking control of the management of debris spinning in space around the Earth may resemble a task worthy of a demigod but is essential. If it is made difficult by the extent of space and time involved, the actual concentration of space capacities in the hands of national agencies and some operators should facilitate the establishment and monitoring of rules of good practice; but for how long? Following the brief constellations fad and in view of the sometimes unexpected development of some terrestrial technologies, no-one today dares to make a real forecast. This is no excuse for not giving a helping hand to Hercules.

6.2 If the sky falls on our heads?

What purpose therefore do Earth-grazing objects serve?

Here I am referring to asteroids likely to hurtle past at less than 45 million kilometres from Earth. Nowadays, we are seeing approximately 800 new objects of this type listed per year, and are more and more aware of these relics from the formation the solar system, not because they have naturally increased in number over recent decades, but for the simple reason that astronomers have acquired greater observation and trajectory calculation powers during this time. NASA even sets aside a specific budget of four million dollars per year for the Spaceguard programme, a network devoted to the study of these asteroids.

6.2.1 Killer meteorites

The name of this American programme indicates the attitude and the mindset that forms the basis of, or most often dominates, the interest in these celestial bodies that regularly visit areas near the Earth: the fear of seeing one of these again, as in Ensisheim, collide with our planet and, because of its size, cause extensive damage or a total catastrophe. The 40 m-diameter meteorite that struck the region of Tunguska in 1908 devastated two thousand square kilometres of Siberian forest and the diameter of the meteorite that was said to have wiped out the dinosaurs 65 million years ago, was thought to have been a dozen kilometres.

What do we actually know about what experts refer to as NEO (Near Earth Objects) and PHA (Potentially Hazardous Asteroids)? Small asteroids less than 50 m in size are very difficult to identify and their capacity to cause harm is, relatively speaking, limited. Very large asteroids, have already been identified and their trajectories are known. The current extent of our knowledge leads us to believe that none of these NEOs will be a threat to our planet during the coming centuries. Average size asteroids, measuring a few 100 m, have not all been identified. The possibility of meeting one is higher and they consequently have a greater capacity to cause damage. At the end of 2008, 5600 NEOs were listed, among which 967 were of a diameter estimated to be greater than 150 m and could come within less than 7.5 million kilometres of the Earth. As things stand, approximately 90% of objects greater than 1 km in diameter have been discovered. In the distant future, apparently a total of approximately one thousand exist, of which one hundred could come close to Earth.

Wanting to know more about these celestial bodies and the threats they can or could constitute led to the law adopted by the American congress which instructed NASA to list 90% of NEOs larger than 140 m in diameter, by 2020. By then, several hundreds of thousands of NEOs will probably have been identified. Of these, there is a real risk that several thousand will strike the Earth in the next one hundred years. For its part, the COPUOS Scientific and Technical Subcommittee set up a specific Earth-grazing object working group in 1999. ESA dedicated the SSA (Space Situational Awareness) programme not only to space debris and to the surveillance of near space, but also to space meteorology and to Earth-grazing objects. The Russians also instigated their own initiative at the start of 2009. There is therefore no doubt that NEOs now come within the fields of investigation and are the responsibility of the communities of astronomers and astronautics.

6.2.2 Apophis: 2029 and 2036

Let us turn our attentions to the case of Apophis for a moment. The discovery of this 250 m-diameter asteroid at the end of December 2004 was overshadowed by the tsunami that hit Sumatra on Boxing Day, December 26th. The Arecibo radiotelescope was used to refine the orbit of Apophis in January 2005, then in August 2005 and finally in May 2006. We must now wait for the end of 2011 then the end of 2012 before new measurements can be taken to once again refine the whereabouts of its orbit.

As things stand, Apophis will next pass close to Earth on April 13, 2029, at 9:45 pm UTC. For several hours, it will be visible to the naked eye from West Africa and Europe. It will cross the equatorial plane at an altitude of 60,000 km, well above geostationary orbit. Circumventing the Earth at an altitude of 30,000 km, it will speed up, consequently increasing its orbital period. The risk of Apophis striking the Earth during this visit is estimated very near from zero. Taking into account the effect of the Earth on Apophis during its travels, astronomers believe the asteroid will return to the regions of the Earth in 2036, or again in 2037. Its passage in April 2029 is decisive for the future trajectory of Apophis and, consequently, for the assessment of possible impact dates, zones and risks. You should remember that a celestial body of this size could devastate a surface area equivalent to several French departments or trigger a tsunami if it fell into the sea, which is statistically more probable.

6.2.3 Dealing with the threat

Dealing with the Threat to Earth from Asteroids and Comets: under this title, the IAA (International Astronautical Academy) wrote a report in 2009, led by Ivan Bekey.[43] This text does not only present the problem and dangers associated with Earth-grazing objects that travel too close to our planet, or the means and programmes implemented by States to gain further knowledge and determine orbits, but also explores the ways which may prevent and manage the collision of such a body with the Earth.

Damage caused by small NEOs can be equated to that caused by disasters frequently or occasionally experienced by human societies. The IAA's report quotes the work of John A. Cross which evaluates the number of major disasters to which mankind has been subjected between 1990 and 2001 to be 38. These disasters result in over 5000 and up to 135,000 deaths (the tsunami of December 2004 caused the death of more than 200,000 people). It is worthwhile trying to assess, model and simulate the impact of the various sizes of NEO, the consequences for affected areas and the time required for life to return to normal. However, are we in fact capable of grasping the true scale of a disaster that could potentially affect the entire planet, e.g. the equivalent of a nuclear winter? Although the evocation of such a perspective seems to surpass our ability to forecast, it is very real in our minds, stimulating in turn our rationality and our emotions, our fears and our anxieties, our hopes and our human sensitivities. It would thus be futile to hope or to claim to be able to conceal its existence and ignore the threat.

The greatest difference between this and other natural disasters is that the impact of a NEO can in principle be detected several weeks and even several years in advance. So how could we conceive that we might not be able to avoid it or, at least, lessen the consequences? Excluding the opportunity of total destruction, e.g. using a conventional or nuclear bomb (this would require the use of devices faster than current intercontinental ballistic missiles), the question raised the most often is the following: is it possible to alter the trajectory of an asteroid? Theoretically, this would simply require a minor modification of the orbit, a deviation, a deflection that would be slighter the earlier on it is performed, to prevent any risk of collision. Experts foresee two types of operation: fast or slow. Fast operations include deviations caused by a kinetic impact, i.e. by sending a space vessel into the NEO itself, at a sufficient relative speed. This is the method that will be tested and measured by the Don Quixote mission organised and implemented by ESA. This mission will send two spacecraft towards an asteroid. The Sancho spacecraft will be the first to arrive at its target and will carry out the observation and study. It will then be joined by the Hidalgo, which will hit the surface of the asteroid, at very high speed. Sancho can then collect data pertaining to the structure of the asteroid, its behaviour under the effect of the impact and the orbital deviation caused.

The types of slow method are more varied, include landing a spacecraft on the surface of the object and operating a sufficient thrust; creating an additional gravitational force using a spacecraft, without direct contact (known as the "gravity tractor" technique); and modifying the Yarkovsky effect by changing the asteroid's albedo or by putting it partially in the shade of a sun screen.

The Yarkovsky effect

When astronomers attempt to estimate impact risks in the long term and, to do this, to perform orbit extrapolations, they must correctly take into account not only the gravitational effects (which they can do with great precision, to the extent of being able to use this technique to control interplanetary probes), but also the non-gravitational effects and, above all, the Yarkovsky effect.

When a celestial body is lit by the Sun, it receives photons that it then retransmits in the form of thermal radiation. This affects the asteroid's speed of rotation and its orbit, in proportions that depend on several factors: its distance in relation to the Sun, its rotation, its size and its form (the Yarkovsky effect affects bodies with a diameter of less than 20 km), and its thermal properties (particularly the thermal conductivity of its surface). These parameters required to determine the Yarkovsky effect are difficult to observe and measure from the Earth. This is one of the reasons why the knowledge and analysis of the orbits of these bodies remains incomplete and imprecise.

Ivan Bekey's report wisely states than whichever technique is adopted, it can only be considered as part of a campaign made up of several missions, or associated with other techniques: altering the course of NEOs is much too difficult an operation, in terms of knowledge to be acquired and technologies to be mastered, to be a sure success at the first attempt.

6.2.4 From an overactive imagination to awareness and liability

Let us return to the question "what purpose can they serve?" which I asked at the beginning of this chapter dealing with Earth-grazing objects, NEOs and PHAs. This question may seem strange or even inappropriate yet it is one that must clearly be asked. Not only because we humans are in some way the beneficiaries of the massive extinction that was thought to have been caused by a giant asteroid hitting Earth sixty five million years ago, but also due to the possible use of this information that smacks as it does of alarmism. Remember how, ten years ago, the release of the disaster films, *Deep Impact* and *Armageddon*, was followed by several killer asteroid alerts; a perfect opportunity (too perfect?) to give credit to the work of astronomers and to space programmes, all too quickly forgotten by the general public and political powers.

The media and cultural players are no doubt a source of temptation for space authorities, whether they are political, industrial or scientific. Are they not effectively in possession of a symbolic, mythical and fictitious potential as unlimited as space? Why not turn to this idea and use it to increase their influence, their social credibility, or even to instigate fear? The end and the means must not however be confused. The space sector would never be so easily "lead astray", as CNES engineer, Michel Avignon, put it. There is a difference between informing the public about Earth-grazing objects and threats they represent on the one hand, and fuelling the fears or firing the emotions and imaginations of our citizens on the other hand. There is no risk, minor or major, that does not require a decision that is as clear and informed as possible.

The threat that Earth-grazing objects pose to the Earth can provoke a number of different reactions. Fatalistic, even defeatist is the reaction of rather not knowing: come what may! This is a reasonable, or should I say prudent, attitude, not altogether unwise. Yet is it human, I mean "responsible" for a human project, concerned for the future of mankind? I do not think so. This attitude is too similar to that of the prophets of doom or that of burying one's head in the sand: not having enough self-confidence, in the capacities and resources available, to ward off danger or to attempt at least to avoid it, to escape it. Just as fatalistic is the attitude that consists in going by the famous French royal sentence: "After me, the deluge" or the as common saying "Sleep in peace, good people". To me these attitudes are not very worthy of a species described as *Homo sapiens sapiens.*

On the contrary, we must be brave enough to confront reality, to scan and survey the sky, as we have always done in the field of meteorology, health, medication, etc. Far from curbing the action or venture, watching is a way of preparing for the future. The more we know, the better we can react. The attitude that consists in listing and monitoring Earth grazing objects to better understand the threats and to, where possible, prevent the less serious of these, is as understandable but more optimistic. How can the attitude that involves opting to invest in the development of techniques for deviation, or even destruction, be qualified? Can this also be described as understandable or does it become unreasonable due to a lack of caution, to too much confidence or human pretension? How is it possible not to fear a misuse of missiles intended to destroy threatening near Earth objects? That said, would succeeding in rerouting an Apophis class NEO, at huge technological and economically expense, not be of extraordinary benefit to mankind as a race?[44]

Whatever the case, scientific and political authorities must learn to tackle and to discuss this type of danger. They must attempt to inform the public with objectivity and honesty, without causing immediate panic, without forgetting to consider ancient fears and cultural whisperings, whilst ensuring the credibility of expertise and scientific approach.

Of course, the threat of an asteroid forces us to think in uncommon scales of time and resources. Whilst astronomers evoke events that could take place in the next few decades or later, our societies, our cultures and our decision-making processes nowadays have to function, act and react in ever-decreasing timeframes. Even so, this situation does not separate us from our human condition, does not prevent us from using our most singular capacities, this combination of caution and daring, good sense and folly, ingenuity and blunder. There is a Gallic warrior in all of us.

6.3 UFO message

2007 was notable for two anniversaries: it was 60 years since Kenneth Arnold's "flying saucers" sighting and thirty years since the creation, within CNES, of GEPAN, recently become GEIPAN (see insert). When readers learn that French experts prefer to use the term UAP (Unidentified Aerospace Phenomenon) rather than UFO, they will immediately understand the reason for linking these two anniversaries. These readers will then ask how this matter, which we all know fascinates the media and requires caution on behalf of political and scientific authorities, is associated with ethics? How can it be linked to the thoughts, actions and liability of a space agency?

CNES and UAPs

Created in 1977 within CNES under the name of Groupe d'études des phénomènes aérospatiaux non-identifiés (GEPAN, Unidentified Aerospace Phenomena Research Group), in an attempt to answer the questions of witnesses and observers of such phenomena, this small unit was renamed GEIPAN in September 2005. The extra "I" for Information demonstrates the CNES' desire to take its responsibility in this sensitive domain to inform the public, and in particular French citizens, more seriously than before.

This mission to inform was therefore added to the three missions on which this group was founded: (1) to gather observations from witnesses and check in each case the dependability of the recorded evidence and of the reliability of the witness; (2) to do everything possible, with the help of a network of experts, to find a plausible explanation for these observations; (3) when it seems impossible to find an explanation due to a known phenomenon, to put together a file that is as comprehensive and accurate as possible, in order to allow teams of scientists outside CNES to carry out any further research; (4) to inform the public of the methods of analysis and results of enquiries conducted, with a clear objective of transparency.

(*Continued*)

The French Airforce, Civil Aviation authorities, National Gendarmerie, National Police Force and National Security forces maintain close relations with GEIPAN.

GEIPAN has a website: http://www.geipan.fr, on which its database contains over 6000 witness statements, classified into 1650 cases, systematically published since April 2007.

I will not go over the history of UFO and UAP sightings, or their human history I should say; it involves a wealth of legends and ancient myths, narratives and mediaeval paintings, reports, photographs and contemporary films. I will not dwell either on the objectivity, commitment or quality of communication that experts, research groups and teams of scientists must have when carrying out their work on these phenomena, on these objects, and on people who take an interest in order to defend their extraterrestrial origin or, on the contrary, to refute it. I simply want to propose and support one idea. As I see it, UAPs, UFOs and other strange manifestations associated with the observation of the sky or with its possible inhabitants, primarily concern human nature, its capacities to observe, to feel, to transmit, but also to imagine, to invent and to discover. Any other observation, hypothetical or even certain, in the scientific sense of the term, that these phenomena could offer, without being disregarded, should be for the time being considered as an added value. And what an extraordinary added value it would be if it resulted in us learning that we, the Earthlings, are not the only intelligent life form in our universe or that ours is not the only cosmic biosphere, regardless of the definition given to the notion of life (see below)! However, we have not yet reached that point.

I would like to start by asking whether UAPs have changed anything within our cultures and for the societies of our era. If so, what are the effects? In his book *Ein moderner Mythus* (Flying Saucers: A Modern Myth of Things Seen in the Skies, 1958), Carl Jung gives his own answer to this question. Having referred to Fred Hoyle's novel *The Black Cloud*, he writes: "Nothing is learnt of the contents from the other side. The encounter with the unconscious edns bootkessly. Our knowledge is not enriched; on this point we remain where we were before the catastrophe".[45] According to Jung, UAPs and the rumours they cause are first and foremost indicative of the human subconscious, the emotional stress and collective plight with which it may be confronted; and the sky, mirror of the human soul, probably heightens this type of feeling. Yet, even reduced to strictly psychological phenomena (a reduction which I do not believe it is possible to accept today), UAPs do not easily convey a message.

Readers may ask themselves why I would insist that UAPs carry or convey any message. Should I pretend to ignore the witness statements and beliefs, the calls and the warnings, the sermons and the prayers that have accompanied the UFO phenomena for more than half a century? Clearly not. Yet are those that we can know about on the same scale as occurrences described and hypotheses put forward? Let us not forget that the question of extraterrestrial life is as old as philosophy: even our ancestors wondered about the possible plurality of inhabited worlds.[46] Do UAPs not offer a definitive response? Apparently not. Whilst Frank Drake's equation is the subject of renewed interest following the development of astronomical and astrobiological research, Enrico Fermi's paradox is also still relevant. To his question: "If extraterrestrials exist, then where are they?", the answers given are far from reaching a general consensus. Between the Drake equation and the Fermi paradox, the solution of betting may remain the most common choice!

Many researchers and curious observers lean towards a second question: why did UFOs and UAPs (at least that which we commonly refer to as such) "appear" in the middle of the 20th century? More rational thinkers like to point out that these phenomena occurred in the sky after they appeared in sci-fi culture, thus supposing a link between the two. To which it can be answered that words and images are needed to talk about and describe what had already been observed; simply remember that, up to the end of the 20th century, the very idea of a flying machine was still only a concept. In reality, UAPs belong to an era marked by science, including if we consider the forerunning sightings from the 19th century. I believe that science fiction is therefore not solely responsible for the appearance, if not of UFOs, at least stories about them, or even a movement called ufology: it was science itself that was requisitioned, that science behind aeronautical and astronautical technologies, that science which allowed the conquest of air and space. Requisitioned but not subjugated: neither supporters of the extraterrestrial original of UAPs, or their critics can claim to possess, through science, definitive proof that they are respectively right. Scientific insight on these phenomena can reveal plenty of surprises and, far from clarifying them with raw evidence, it can on the contrary leave them tinted with interrogation, imagination, doubt and finally faith.

The third question raised by UAPs to those who accept that they may be subject to ethical interrogation is that of otherness: for us, as 21st century humans, what "others" could these phenomena reveal or conceal? What "others" could these flying objects be transporting? "Others" who impose their presence along with their difference? "Others" who can be imagined, desired, repressed, ignored, annihilated, sometimes tutelary figures, sometimes threatening shadows? The *alter,* the alien, the stranger can become an *alter ego,* an ally, a friend who alone would know how to end the dreadful feeling of solitude which sometimes clenches the human heart. It can also reveal itself in a less radiant light as the bearer of conquering, aggressive intentions. Beneficial or baleful, auspicious or unlucky, the appearance of the alien is at the root of the craziest but also the most effective hopes or fears that mankind can develop. The other is a permanent source of fascination for mankind to the extent that we seem to struggle to live without its existence, without its presence, whether real or imaginary. It is not therefore at all surprising that the slightest appearance of another being, proven or hypothetical, identified or unidentified, provokes those reactions that we know so well. And if it were true, if each UAP, each UFO was our chance to discover another being, to meet them, would the existence of each of us not be dramatically changed? It has to be admitted that more often than not, nothing much comes of such sightings. We mostly just let out a sigh, or hear it from someone else: "If only the evidence was more conclusive . . .".

Yet the truth remains concealed and avoids all attempts to understand it completely, to subject it to proof and evidence: rather than proof, we find ourselves faced with tests. Tests for human intelligence and imagination: the scientific approach and more still human awareness are condemned to be pulled, torn between the known and the unknown, between knowledge and ignorance. Tests also for our societies and our cultures, even if, as I have said before, they have scarcely changed as a result . . . .

Faith may be one way of distancing and avoiding this test. Answers are given that are as complete and as global as is possible, not to say fundamentalist, in response to all questions and indecisions. To think, believe or know otherwise is deemed impossible, unorthodox and heretical. The pyres of the Inquisition, stocks and iron shackles have not all disappeared; they have simply been replaced by common law courts or politically, scientifically, religiously or "ufologically" correct sectarians! The scientific and religious syncretism that marks a number of movements founded in the belief in flying saucers, can probably be explained by the presence and influence of these processes of belief or even, to use an expression from Wiktor Stoczkowski, of "intimate conviction."[47] Ignorance is perceived as a fault, a flaw that needs to be eradicated or used for one's own ends. Yet, if mankind likes to be called by the impressive title *Homo sapiens sapiens*, to claim to be aware of itself and of its own conscience, it must also recognise the limits of this conscience, of this knowledge and admit its ignorance. In 1440, Nicholas of Cusa published a work which he entitled *De Docta Ignorantia* (Learned ignorance): this philosopher and clergyman believed that humans were only as wise as they knew themselves to be ignorant. A good lesson.

When dealing with UAP and UFO sightings, experts are therefore often forced to a halt through an admission of ignorance. There are ways to get out of this intolerable deadlock: to mount the horse of evidence and dogmatism, to pretend to ignore the causes of this ignorance, etc. These ways are not acceptable: they overlook the singular experience of witnesses by forcibly imposing limits and the forms of certainty and correctness, often using an ideological shoehorn. Progress must be made in terms of the technological and scientific handling of UAPs, in order to come within reach of the truth or the reality they represent: but, at the same time, efforts must be made to listen to the witnesses themselves. We simply need to consider the extraordinary variety of UAPs recorded in the CNES archives. There are probably many resemblances, analogies and constants yet what of the immediately apparent or more subtle originalities and singularities? Are they the sole result of phenomena themselves or also that of observers, Earthling witnesses? Do they not reveal something human that we have not yet come across or have not yet fully understood?

The contemplation of the sky and the conquest of space are not simply daydreams of no importance or nostalgic thoughts. On the contrary, they require commitment from individuals and society at large, even in their most human and most terrestrial provinces. It is for this reason that they deserve to be or even must be accompanied by ethical thinking.

[43] *Dealing with the Threat to Earth from Asteroids and Comets*. Ed. Ivan Bekey. Paris: International Astronautical Academy, 2009.
[44] See Jamet, Didier & Mottez, Fabrice. *2012. Scénarios pour une fin du monde*. Paris: Belin, 2009.
[45] Jung, Carl Gustav. *Flying Saucers. A Modern Myth of Things Seen in the Skies*. Princeton (NJ): Princeton University Press, 1978: 125.
[46] See Dick, Steven J. *Plurality of Worlds. The Origin of the Extraterrestrial Life Debate from Democritus to Kant*. Cambridge: Cambridge University Press, 1982. Crowe, Michael J. *The Extraterrestrial Life Debate, 1750–1900*. Mineola (NY): Dover Publications, 1999.
[47] See Stoczkowski, Wiktor. *Des hommes, des dieux et des extraterrestres. Ethnologie d'une croyance moderne*. Paris: Flammarion, 1999.

CHAPTER 7

7 The Greater Earth

"Is our modern world ready to accept that some spaces and some resources cannot be nationally appropriated? It would seem so. Is it ready to think about subjecting the economic gains of the exploitation of these resources to a universal democratic process by ignoring the differences in the technological and financial levels of members of the international community? As international relations stand, it would seem not."[48]

Space is no longer exclusive to the gods as Ptolemy already suggested: "But when I trace at my pleasure the windings to and fro of the heavenly bodies, I no longer touch Earth with my feet. I stand in the presence of Zeus himself and take my fill of ambrosia". Humanity may have succeeded in reaching space, but does that confer a right to ownership? Must it, can it apply the thoughts, methods and arrangements to which humankind and its societies have traditionally resorted on Earth? When all is said and done, to whom does space belong?

7.1 Space, the heritage and province of all mankind

The forefathers of space law, the policy leaders who committed their countries to signing and ratifying space resolutions, agreements and treaties put forward and supported an essential notion, that of the freedom of exploration and use, with the association of the notions of non-appropriation, cooperation and the prohibition of some activities (e.g. in terms of nuclear weapons). To do this, they resorted to several legal notions, in particular those of common property, common heritage and province.

The first article to the Outer Space Treaty defines the foundations of the status of space: "Article I: § 1. The exploration and use of outer space, including the Moon and other celestial bodies, shall be carried out for the benefit and in the interests of all countries, irrespective of their degree of economic or scientific development, and shall be the province of all mankind. § 2. Outer space, including the Moon and other celestial bodies, shall be free for exploration and use by all States without discrimination of any kind, on a basis of equality and in accordance with international law, and there shall be free access to all areas of celestial bodies. § 3. There shall be freedom of scientific investigation in outer space, including the Moon and other celestial bodies, and States shall facilitate and encourage international cooperation in such investigation."[49]

What does the 1967 Treaty claim in this first article? Basically, that space is not a *no man's land* that could be occupied and exploited by the first to arrive, or by any subsequent parties. The Treaty defines space as having an international character and cannot be claimed or appropriated by any State. It must be considered as a public domain where everyone must ensure order, peace and equality between States. Making a statement like this in the mid 1960s was clearly significant. Thus, at a time when the two largest world powers, the US and the Soviet Union, were going head to head in a race to the Moon, in the context of the Cold War, this Treaty proposed the protection of space against any national claim and any military hankering, offering it equally to all mankind, no less. The statement was generous, prophetic and even utopian. It had the great merit of being adopted, supported and confirmed by States, of a varying number according to the texts proposed for their signature. The Outer Space Treaty was ratified by 97 States and signed by 27 others. The Moon Agreement was only ratified by 12 States and signed by 4 others. So, what status does space law grant to space?

The first article of the 1967 Treaty has the underlying legal idea of "common heritage". It describes what is stated to belong to everyone and must consequently be used for the common good, with everyone's share (e.g. of resources) being guaranteed. The 1960s gave rise to the concept of the common heritage of mankind against a backdrop of decolonisation, the bipolarisation of international relations and the emergence of new North-South relations. It is also linked to the notion of "new international economic order" claimed at the same time by newly decolonised States, based on equity, interdependence, sovereign equality and interstate cooperation. Sea has been the first domain where this notion was used. Following the proposal made by the Maltese ambassador Arvid Pardo in 1967, the international Montego Bay Convention, signed in December 1982, returned to the same concept of the common heritage of mankind to apply it to the seabed in order to preserve the polymetallic nodules from any commercial exploitation. On the other hand, surface waters, when they are found in international zones, are treated as "common property" (*res communis*): they belong to no-one and are therefore accessible, usable and can be exploited by all, possibly subject to some restrictions or regulations; no State can claim any form of sovereignty over them and no national law applies to them, although international agreements do provide certain national measures and the enforcement of fishing quotas.

On closer inspection, space law also refers to the concept of common property. Whilst the first article of the Outer Space Treaty apparently accords space the status of common heritage of mankind, the second article refers to it more in terms of *res communis*: "Article II: Outer space, including the Moon and other celestial bodies, is not subject to national appropriation by claim of sovereignty, by means of use or occupation, or by any other means". The resultant difficulty clearly lies in the introduction of the notion of "province" (in French, *apanage*), in the first article of the Treaty: the exploration and use of space are qualified as the province of all mankind.

What does "province" mean? Article IV of the 1979 Moon Agreement also refers to this notion: "The exploration and use of the Moon shall be the province of all mankind and shall be carried out for the benefit and in the interests of all countries, irrespective of their degree of economic or scientific development. Due regard shall be paid to the interests of present and future generations as well as to the need to promote higher standards of living and conditions of economic and social progress and development in accordance with the Charter of the United Nations". During the era of royalty in France, the term province referred to the share of the royal kingdom granted to the younger sons of the royal family in compensation for their exclusion from the throne. Since then, the term has become more generalised, meaning property, inheritance. It retains a notion of elitism.

The use of the notion of province, to refer to space is interesting. On the one hand, it offers mankind a rightful position: not that of domination (mankind does not rule the universe), or that of submission (no less deserving of its share), but more that of an heir benefiting from the work, power and potential troubles of its predecessors, who would like to be able to provide its descendants with the same possibility. On the other hand, still referring to space law, what mankind claims as its inheritance is not actually a territory, but a mission: that of exploiting and using a portion of the universe (the Moon in this case), for its own interests and that of future generations. However, does the idea of province not jeopardise the idea of a cultural and environmental protection of space? Does this not take us back to the status of public property? The example of the exploitation of marine fishery resources, as practiced today, can justify a certain perplexity. Thus, the law is faced with the innate difficulty of defining space itself.

In the first chapter, we saw that this issue is one of contention. Opposed to a "solid", physical and topographical space, is a space formed by moving bodies, not only of natural origin but now also of human origin, and even by human beings: a set of trajectories and movements, knowledge and techniques and, now, exchanges and relations. The following words from Jean-Jacques Salomon about technological objects can be applied to space: "industrial machinery – 'machines reproduced by machines' – moved us away from the natural history of the tool, whose limit was biological, to enter the new era of the artificial history of technological objects whose limit is no longer determined by the limit of our knowledge alone".[50]

In short, as I previously indicated, space is not only a place, but also what humans do there and what they do with it. Its limits are not primarily geographical or natural, but mostly those of our knowledge and our scientific and technical ignorance. Space law defines the place as a common property or common heritage and the activities (exploration and use) as a province. Are these two notions compatible? Is it not important to recognise that any use is a form of appropriation, even temporary, applying to things ranging from software which can be used even without user rights, up to the surface of the oceans travelled by human vessels? Space is not exempt from this trend, or this risk even, as can be seen by the example of geostationary orbit.

In the mid-1970s, certain equatorial countries (including Brazil, Colombia, The Congo, Indonesia, Uganda, Kenya and Zaire) claimed rights to the portions of the geostationary orbit located above their territories; in 1976, they even signed the Bogota Declaration asserting their sovereignty. As I mentioned before, this orbit, located 35,800 km from the Earth and vertically to the Equator, has the specific nature of being in phase with the Earth's rotation; placed in this orbit, a satellite appears practically immobile in relation to area of the globe over which it is found. This characteristic of geostationary orbit is useful in terms of telecommunication satellites which, in this situation, can ensure an uninterrupted transmission to a determined portion of the Earth's surface; more than 250 satellites currently occupy this orbit. As a result of the Bogota Declaration, the International Telecommunications Union (ITU) took charge of the allocation and allotment of the places along the geostationary orbit, therefore averting the attempt to claim ownership of this part of outer space by the countries located below the orbit and along the Equator, based on geographical criteria. Even so, is this a true or complete application of the principle of non-appropriation? Are satellites not a form of appropriation of a position, of a portion of orbit, beyond even their life span and their service life, as they may eventually continue to occupy this position as debris? Here the aforementioned notion of launch State emerges as another example of the shift towards the operational criterion: the launch State which brings about or authorises launch operations is legally liable for the consequences that could ensue. To what extent is it possible to refuse this State any form or at least any idea of appropriation, of an orbit for example? So, to talk of the non-appropriation of space, may be utopian, as space has already entered the era of operations, usage and utility, to such an extent that we now have to consider situations of multi-use and multi-appropriation.

For the record, the planting of American flags on the Moon did not constitute a national appropriation of the lunar ground but was simply a symbolic act. Moreover, as of 18 November 1969, the United States Congress adopted a law directed at NASA, prohibiting this type of act or reducing it to a simple act of national pride. In the boxed text below, I give an example, of course terrestrial, of the multiple uses and appropriations of space objects, here for liturgical and museographical purposes, which simultaneously introduces the difficulty of cultural diversity in the understanding and use of cosmic, space actualities.

Joint custody of a sacred meteorite by a museum and an American Indian tribe

"Following a drawn-out legal battle, the American Museum of Natural History in New York and an Indian tribe from the State of Oregon (North West) agreed to share a meteorite weighing over 15 tonnes. The space rock baptised 'Willamette' by the museum and 'Tomanowos' by the Clackama Indians, who regarded it as sacred, will continue to reside at the New York museum. However, the Oregon tribes, members of the Confederate Tribes of the Grand Ronde Community, have the right to visit it. 'The Confederation will be able to renew and maintain its relations with the meteorite' stated the museum's director Hellen Futter. According to scientists, the celestial fragment was thought to have fallen to Earth over 10,000 years ago, near the Pacific, and to have been pushed by a glacier where it ended its journey in the Willamette Valley, in Oregon. The Clackamas had issued a request in February for the return of the meteorite that they considered to be sacred and which they used during their rites. Both parties claimed to be satisfied with the agreement. The museum will continue to exhibit the enormous rock, as it has done since 1906, and the Clackamas will be able to hold their purification ceremonies at its base."

(AFP dispatch, New York, 23 June 2000)

To the criteria of use (*usus*) must also be added to the criteria of product (*fructus*). The use of space can in effect take the form of exploitation, in other words lead to the extraction or production of tangible or intangible assets, followed by its valuation, not forgetting the question of any by-products. What status should be given to the products of space and any by-products, drawn from the exploitation of common property or common heritage? Can they be subject to exclusive use? The answer is simply and definitely yes, as illustrated in the remote sensing sector. Information resources, gathered by satellites then processed by teams on the ground, become the property of those who have the requisite technological and financial means. As soon as we start wondering about the future of space, we are entitled to ask whether the influence of the market will become predominant in the decisions taken, including with regards to legal positions and ethical opinions. If engineers found a way to exploit the polymetallic nodules or any other seabed resource, in raw materials or energy, in accordance with reasonable, or even advantageous technological and economic terms and conditions, what would happen to the status of common heritage of mankind granted to the seabed by Montego Bay Convention? Similarly, if lunar resources did start to be exploited or a competitive market were to appear, would space law not be in danger of becoming powerless, faced with economic issues? This is already the case in the communications sector in which space technologies challenge the basis of State monopoly. In effect, they dematerialise telecommunication networks and render the development of the public domain pointless. "The satellite, private property, is placed in a space over which there is no national sovereignty. The receiving antenna is on privately owned premises. No natural monopoly takes place and there is no longer a need to resort to the public domain."[51] When public power looses or gives up the attributes which it had hitherto been conferred, there is a risk that these attributes will quickly appear on the market. Eldorados, gold rushes, with their share of excess and exactions, will never definitively be consigned to the past.

7.2 Is the Moon for sale?

Asking for the Moon is, in many cultures, a common symbol of supreme and ultimate desire. Could mankind have something in common with the wolves which, since time immemorial, have howled at the sight of the silvery orb? Whatever the case, the absence of reference to private parties in the 1967 Treaty has made it possible for individuals and companies to come along with the answer to this question and actually offer lunar plots for sale.

The first lunar real estate undertakings date back a long way. In the history on the subject by author Virgiliu Pop, the story begins in 1756, when Frederick the Great, King of Prussia, allegedly gave the Moon to his subject Aul Jürgens as a reward. Then, Pop explains, starting in 1937, a steady stream of ambassadors and vendors began claiming stakes on the Moon and selling it in plots of several hectares or acres. Of these, Dennis Hope, CEO of the Galactic Government and Head of the Lunar Embassy, is today the most active.[52] These Moon salesmen would use any argument to defend the genuineness of their trade with the argument "first come (or first recorded), first served" probably the most common. There are others too: a federal American law from 1862 allowing any individual to claim land not yet belonging to an owner; the purchase of a Lunakhod, left on the surface of the Moon by the Soviets, giving rise to the opportunity to claim ownership of the surrounding terrain; young Romanians deciding to finance the construction of a sports centre by selling plots on Mars, as celestial bodies other

Lunar Land Claim
Certificate of Registration

Lunar Properties LLC™, for and in consideration of good and valuable consideration to them paid by grantee, receipt of which is hereby acknowledged, do hereby grant, bargain, quitclaim and convey onto said Grantee:

Your Name Here

and his/her heirs now and forever, the following Lunar Claim Registration
located on Earth's only Moon, Luna:
BONPLAND crater area, located in Mare Cognitum, "Sea that has become known",
17.4° West Longitude, 8.3° South Latitude
A 1 hectare Registered Lunar Land claim, Locator Number: BONPLAND-2007-76037-01-001-001
Subject to the terms and conditions as set forth by Lunar Properties LLC
Registered with the International Celestial Lands Registry™ by Lunar Properties LLC™,
on this 31st day of December 2007

Deed Number: 04261968

Legal Recorder of Lunar land claims for Lunar Properties LLC™

Fig. 6. *A Lunar Embassy Deed (source: Internet).*

than the Moon had also started to be sold, etc. Some "ambassadors" issued the Apollo mission astronauts with passports so that they can legitimately set foot on the Moon's surface, whilst others demanded that NASA pays a tax to gain authorisation to land lunar modules, LEMs, on Selena's surface. Is selling the Moon profitable? This is of course very difficult to say, even if Hope's Lunar Embassy claims to have sold more than one and a half million hectares at 88 euros a piece: Where then do we draw the line between farce, as some vendors call it, and intellectual fraud?

Legal experts are categorical when questioned: these commercial activities are totally illegal. If the Outer Space Treaty does not mention private individuals, it is primarily because, when it was drawn up, only States had the technological and financial capacity to undertake the exploration and exploitation of space. Even so, it would seem logical to apply the prohibition of ownership, not only to the first parties in space, but also to all those who follow, without exception. As to the property titles established under American legislation, they obviously have no legal bearing at the international level.

Can we therefore conclude that legally, space can definitively be considered as a *res extra commercium*, a thing outside commerce, which may not be the property of an individual nor the object of trade, such as the human body in French law? It is perhaps not as simple as that when we consider any resources extracted from celestial bodies. As we have seen, the notion of common property deems the resources freely accessible, and the common heritage principle means they must be shared equitably. The sixth article of the Moon Agreement grants States "the right to collect on and remove from the Moon samples of its mineral and other substances. Such samples shall remain at the disposal of those States Parties which caused them to be collected and may be used by them for scientific purposes." Even so, "States Parties shall have regard to the desirability of making a portion of such samples available to other interested States Parties and the international scientific community for scientific investigation". So be it, but who will monitor these operations and, if necessary, have the capacity to sanction them? How will the shift from the field of science to that of commercial applications or exploitation be managed? Many legal experts have wondered about the current situation. As things stand, the non-ratification of the Moon Agreement by the main space powers creates a true legal vacuum. So, is it necessary to continue with the current international regime of resource exploitation, even if it means amending it in favour of space powers or establishing it beyond the context of the agreement? Or would it be more advisable to move away from the current provisions of space law, and to reinterpret the principle of non-appropriation by leaving each State free to apply its own national legislation? The Moon may never be put up for sale but it is already in the hands of those other than scientists, and is no longer in the sights of poets alone.

7.3 Space business or economy?

At the start of the 21st century, our planet Earth is taking stock of the dramatic and disastrous consequences which may result from an over utilitarian and productivist approach: devastating pollution, toxic waste too easily ignored, global climate change, etc. If there is hope of seeing all social, economic, political and scientific players reach a consensus concerning the measures intended if not to stop, then at least to react to these phenomena, we have to admit that the reality of this consensus and its date of implementation do not inspire a widely shared optimism. Does the future of space have to be subject to a similar perspective? The growing place occupied by the utilitarian dimension, and the role of industrial and financial companies, more or less independent of States, provoke thought, or even fear.

There are three possible approaches. That of absolute sanctuarisation would involve ensuring not only integral protection for space, but also the recognition of a transcendent character, beyond all current economic calculations. If this attitude is not one with which our era's commercial civilisations and so-called developed civilisations are familiar, it is without doubt familiar to other modern day and future peoples, whose rights are perfectly equal to our own. Opting for controlled exploitation (the second possible approach) would be based on specifications to govern all forms of exploitation. This would need to be written beforehand and not "on the spot" or during a crisis, as is often the case on Earth. Finally, the "first come, first served" approach would involve adhering to the now dominant rules governing exploitation and economic efficiency.

Sanctuarisation, i.e. the integral freeze of celestial bodies that mankind sees as gradually entering its domestic universe, can only be a intermediary solution, by default, which can however be respected over a long period of time; once a use with high economic value or a strategic use arises, the pressure for its practical implementation will be irresistible. On the other hand, market forces alone would not be sufficient to hold the long-term interests of mankind. I will take the example of celestial bodies. If they are willingly part of exploitation projects (even if the resources they have remain limited), it should not be forgotten that they carry memories, physical evidence dating back billions of years. Finally, what form could a controlled exploitation take? Based on a status of *res communis*, the community of nations would be in a position to concede to an eventual private or public requesting party, a right of exploitation of this public resource for a given project and a given time; specifications would need to be drawn up prior to this to guarantee the acceptability of the planned operations. Based on the hypothesis that substantial economic prospects could be unearthed on nearby celestial bodies, it would be ideal for all humankind to profit from these globally and for the emergence of a new resource to be really embraced as a good thing for all of humanity.

Is thus seeking and applying standards that would ensure the smooth running of our "Earth home" (*oikos*) not simply about applying economic virtues (*oikos nomos*: the law governing the home)? From this point of view, space in the future would thus need to become truly economic, without however falling into the hands of the market. The momentum of the market is very useful in establishing and maintaining an economy but its logic is too restrictive for steering and managing society. With regards to the use to be made of our new global village, community village, the next decades will need to define management or a policy that controls the market, respecting people and human societies, in particular their diversities. This village could tomorrow become known as *Greater Earth*.

Greater Earth defines the area, the space territory that surrounds the Earth and where most future space activities could take place, without requiring overly high energy costs. Around our planet, there is in fact a roughly spherical area with a diameter that exceeds a million kilometres and in which a space platform is either naturally held by the Earth's gravity, or maintained in this terrestrial neighbourhood at the cost of modest propulsion manoeuvres. The L3 Lagrangian point in the Sun-Earth system is one example. Platforms and bases developing in this domain, instead of being at the bottom of gravitational potential wells, can be easily accessed: from the Earth, the propulsive evaluation is, within 1 or 2 km/sec, the same as that of placing into geostationary orbit. The propagation time of radio-electric waves does not exceed 3–4 sec, signifying that exchanges with the Earth may be naturally continual and cooperative, without being autonomous but only controlled from time to time. Finally, in the event of human presence on board, the time required to return to Earth is a few days, comparable to that of lunar missions, whilst being cheaper in terms of propellants.

Experts brought together by ESA at the start of the 2000s, see the next logical step for space venture, for the reasonable use of space by our civilisation as the potential development of this Greater Earth. Initially for reasons of usefulness, and then because it is sensible in terms of the management of technological systems, and finally because mankind will eventually play a useful role in these systems, a logical step will not crush missions with its own constraints: a large number of missions can be run in telepresence mode. Jacques Breton writes "Fragile and vulnerable, Greater Earth is the space domain linked to the Earth and its inhabitants. It holds a lot of memories and bears traces of the history of this small area of the universe in which life appeared. It is important to know how to anticipate and implement ethical questioning so that man again becomes involved in these debates. As Edgar Morin remarks, the idea of 'Homeland Earth' constitutes the only answer to internationalisation and, conversely, it makes us human".[53]

It is advisable to establish and implement an economy of the growing space sphere of Greater Earth. To do this, it is not necessary to wait for a market to be established there. On the contrary, as long as this space maintains the status of potential or virtual, it would be advisable to encourage, to promote values, forms or expressions of a space economy. What would be the foundations of this? Some people advocate cooperation, concern for poorer populations. Why not? However one of the questions, among many others, that needs to be asked about the space domain is: to what structure can the establishment, implementation, management and monitoring of such an economy be entrusted? There again, space finds itself faced with the question of sovereignty, primarily, that of States. Must these activities "today be considered by a State, such as France (or a regional entity, such as the EU), as an indispensable part of its sovereignty?" When thinking in these terms, by sovereignty, Roger Lesgards means the "capacity to be truly autonomous, only relying on itself to ensure its own safety, defence, interests and even its political system and status as an international power".[54] His response is the following: "Yes, space activities are very much involved in the military and economic elements of French and European sovereignty".[55] Space is effectively associated with a desire for power, domination, in particular in the two domains of the economy and military, to the extent that none of the countries that have embarked upon the space venture, has yet totally renounced it. On the contrary, new countries, even among those who are deemed to be developing, are looking to participate. Even though the idea of power is now demonstrated as much by participation as by opposition or conflict, nevertheless, the two military and economic fields remain very coveted. Is the idea of a planetary space economy utopian? It is easy to be extremely pessimistic here, but why not take seriously the increasing role, beside international and intergovernmental organisations, of those so-called non-governmental organisations? Let us be realistic: the creation of a space NGO is probably not imminent, but we should not underestimate the influence of public opinion, however vague this notion may be. The relevance of this opinion, if I can express myself in this way, is in asking questions both in terms of credibility and in terms of worth. Despite the trivialisation which affects it and the fascination it arouses, space still appears to be, in terms of public opinion, at the stage of seeking credibility.

And what if the new idea, "more powerful than any other in history" predicted by Fred Hoyle was something like this Spaceship Earth, even this Greater Earth? Not devoid of dreams or utopia, this new idea immerses its roots in the concreteness of a scientific and technological, economic and political, cultural and ethical venture of which we humans do not need to be embarrassed. If, by doing so we prove Edmund Husserl right, do not however forget to consider the ancient impetus that inspired, forged, moulded our species, that of explorer; however, we will not neglect good old Mother Earth, "the good Earth", according to the words of Franck Bormann, from lunar orbit, at Christmas 1968.

[48] Voeckel, Michel, 1990. Quoted in Akbar, Sabine. "La lune, patrimoine commun de l'humanité?". *IFRI memo*, 2006.
[49] For all treaties, agreements, etc., visit http://www.oosa.unvienna.org/oosa/SpaceLaw.
[50] Salomon, Jean-Jacques. *Survivre à la science. Une certaine idée du futur*. Paris: Albin Michel, 2000: 43.
[51] Lebeau, André. *L'Espace. Les enjeux et les mythes*. Paris: Hachette, 1998: 140.
[52] See Pop, Virgiliu. *Unreal Estate. The Men who Sold the Moon*. www.virgiliu.com, 2006.
[53] Quoted in Arnould, Jacques. *La Seconde Chance d'Icare*. Paris: Cerf, 2001: 149.
[54] Lesgards, Roger. *Conquête spatiale et démocratie*. Paris: Presses de Sciences Po, 1998: 89.
[55] *Ibid.* 93.

CHAPTER 8

8 Exploration

"Together let us explore the stars." These words spoken by John F. Kennedy in 1961 are still relevant today. Even if many nations recognise that the continuation of the space venture, particularly towards deep space, is not possible without increased international cooperation, the reality of such cooperation is still a far cry from the vision of the American president. His invitation to explore the stars remains however both a project and a responsibility claimed by a number of countries, regardless of their more deeply rooted motivations or associated reasons: competition and the demonstration of a technical, scientific and political ambition are some of the main reasons. We must not allow these reasons to obliterate the biological and human roots of inclination, propensity and enthusiasm for exploration, which has driven the space venture from the outset.

8.1 Space, so human

As strange and foreign as space can be or appears to be, as a location or an activity, it is nevertheless founded on characteristically human elements: I like to call it drive, as we are talking about an action and a responsibility.

8.1.1 Explore, conquer, flee

Is space about exploration or conquest? I am often asked this question, which implies a difference between these two terms. The first, exploration, is readily given a peaceful meaning and connotation and the project, the attempt to expand knowledge that mankind has not only of itself but also of regions, lands, seas and now space. The second term, conquest, designates the propensity to occupy new lands, to establish oneself there, to develop economic and industrial activities, to exploit their resources and to obtain new resources. Until now, the space venture seems to have readily pursued and structured both of these approaches together. The probes which fly over the planets of our solar system or the on-board instruments which scan the universe are simply exploring, whereas the networks woven by satellites around the Earth participate in the conquest of this area of near space.

Whilst all species, whether animal or plant, continually try to expand their territories and to conquer new ones, our species has the distinct ability to be able to freely leave its territories, the places where it normally lives with no immediate desire for conquest. The role of the imaginary, of the immaterial in the history of mankind consolidates this idea: isn't it all about forms of exploration and intellectual appropriation, original, non destructive and, all things considered, non conquering? Even so, it is essential to recognise that pure exploration, which involves "having a look" for curiosity's sake and leaving again, often only lasts for a limited time and has often led to, as the history books show, more conquering ventures.

Whatever is true of this distinction which I felt I had to make, the human species demonstrated very early on in its history a unique propensity for exploration, conquest and migration, as if driven by a congenital curiosity, which mankind's specific culture has only served to reinforce, to the point of sometimes giving mankind easy excuses. This is what the maxim of 17th century French author, François de La Rochefoucauld claimed: "Crimes are made innocent, even just, by their number and nature; hence public robbery becomes a skilful achievement and wrongful requisition of a province is called conquest". As does this sentence from another French author, Bernard de Fontenelle, in his work *Entretiens sur la pluralité des mondes* (Conversations on the plurality of Worlds, 1686): "The only thing men experience in the pleasure of being loved is triumph over the person who loves them; and happy lovers are only happy because they have conquered". The human condition does not easily forget its biological origins.

Be that as it may, very few, if any, other species have an area of expansion as vast as our own, in such diverse and varied conditions. This is surely the consequence of our capacities to adapt to these multiple environments and to adapt them to our own requirements, the consequence also of our cultures and our traditions. René Dubos, whose name I have already mentioned, devoted a great deal of his thinking to this very human characteristic.[56] Even so, we cannot ignore this "urgency to explore" which supposedly transformed, according to Ben Finney, the juvenile characteristic of curiosity, into an adult profession.[57] We are all would-be Christopher Columbus's or Captain Cooks: the large waves of migration of human populations, including those of modern times, from the European continent and those which occurred between the Oceania islands, were not only be the result of economic and demographic constraints, but also the expression of a drive profoundly rooted in our nature. Without doubt, when President Thomas Jefferson raised the idea of an expedition to the west of Mississippi, which Lewis and Clark undertook between 1804 and 1806, his reasons were both political (the *Declaration of Independence* had been in existence for thirty years) and economic (in order to expand US trade). Moreover, President John F. Kennedy, who called for international cooperation for space exploration in 1961, was not unaware that space already constituted a new field of competition between the major powers.

Political or economic benefits may not be the only reason, the only alibi: to instigate and continue such movements, to stop them from dying out due to lack of interest and even (some say "through lack of passion"), it is necessary to refer to or to admit the existence of a more profound driving force, mythical I would say to hark back to the story of Icarus, which is also based on the tendency towards expansion that marks living beings, to implement with unusual brilliance, the drive of humanity, which combines awareness and intelligence, imagination, fear, hope, culture and art.

Describing man as explorer and conqueror, as a migratory being, must not lead us to forget that man is also sedentary; a condition that was essential in the emergence of agriculture, cities and sometimes even culture. The mythical tales of Ulysses or Jason and the Argonauts are there to remind us of the happiness we feel on returning home or coming back to one's land, in order to pursue or live out one's existence. Aware of their slim chances of survival, the three American astronauts who took part in the dramatic Apollo XIII mission, supposedly asked to die as close as possible to Earth, if they could not be saved. However, the issues of the relationship between nomadism and sedentarism, between distant exploration and attachment to Homeland Earth, will be even more apparent if missions to Mars, involving great distances and long periods of time, one day become a reality.

The act of fleeing, whether admitted or not, is exempt from this latter aspect and is even its main characteristic: it thus involves leaving a place, never to return, for somewhere better if possible. Thus, the passengers on board the Mayflower, like other migrants motivated by their religious convictions, left their homeland, convinced they would find the lost Paradise, the Garden of Eden of their ancient ancestors, Adam and Eve, on the other side of the Atlantic, where they would at least be able to set themselves up a New World.

Let us not forget that fleeing was the essential reason for the mythical tale of Daedalus and Icarus and, even if achieving this on a planetary scale belongs to a more hypothetical and distant future, it nevertheless is at the centre of the apprehensions of space venture via the most common, most shared imagination. I will use the example of a survey conducted in Belgium, as part of the events associated with the countdown to the new millennium. Ten years ago, the museum of the Royal Belgian Institute of Natural Sciences in Brussels organised an exhibition on the topic: "What will be tomorrow's world?" On this occasion, a survey was conducted on around 500 children, aged between 10 and 13, in several schools in Belgium. They were asked to draw a picture of what they thought planet Earth, nature or their town would be like in 2015. The organisers commented: "Faced with the negative impression they harbour of the status of the planet and its future, explained the organisers, some (22 of the 106 drawings representing the planet in 2015) suggest leaving this now uninhabitable Earth to live on another planet or an interplanetary station, whilst others (45 of the 106 drawings) portray technology (often futuristic) coming to the rescue to restore a more viable environment". For example, one of the drawings represented the Earth "that had just exploded" and several rockets carrying survivors to a space base. This demonstrates, or else proves the persistence of the myth of Icarus, even among young generations. The geographic and cultural extent of this still remains to be seen, however. How far indeed does the awareness, or even obsession, of the limit and scarcity of natural resources stretch beyond modern western societies and their global vision of reality, bound as they are by satisfying their needs? It would be useful to conduct a similar enquiry to this in different cultural contexts, particularly those of other space powers.

Explore, conquer or flee: for various different reasons, space venture appears to be deeply rooted in human cultures, in the understanding that humans have of themselves and of their destiny, in their most profound aspirations. Roger Lesgards talks of "the need that modern man may feel to find, in a sky he now knows to be devoid of a divine presence, in a cosmos where great myths have lost their authority in the light of scientific progress, a renewed form of elevation and transcendence of which he himself would be the creator. Man sees the Space adventure as therefore constituting the beginning and the promise of a new epic".[58] It is not therefore surprising if other ideas, other notions, dear to traditions and cultures, are easily found in space discourse – I am thinking, in particular, of those of the horizon and frontiers.

8.1.2 Horizon, frontier and utopia

One French dictionary defines the horizon as "the apparent intersection of the Earth and sky as seen by an observer. The horizon moves away, retreats and escapes as we try to get nearer". Since it can be observed, the horizon incites scientific interest and the implementation of investigation techniques. At the same time, because it continually moves away from he who tries to approach it in the hope of finding out more about it, the horizon remains just as mysterious as it has ever been, if not more so. It thus has a power to fascinate that has never changed through time. This power of the horizon is thought to have been the origin of most major ventures recorded in human history: the pyramids of Egypt lost to the horizon, the walls of China that block the horizon and protect from strangers and even modern means of transport (transatlantic ships or spaceships) that take us beyond the seas or the sky, in other words beyond what would seem to separate us from the horizon. These ventures belong to what Eugene S. Ferguson referred to as *Macro-Engineering Projects* (MEPs).[59] Ferguson explains that these projects push the boundaries of knowledge and resources of their time. Their realisation is complex, long and expensive and they are often the act of visionaries or at least enthusiasts, rather than leaders concerned by financial constraints, the immediate needs of a company, or the consequences on the latter of such a project! "There are three roads to ruin, said the great Rothschild: gambling, women and engineers. The first two are more agreeable – but the last is most certain", wrote, not without a pinch of irony, August Detoeuf.

In his book, entitled *Les Horizons chimériques*, Roger Bonnet, former scientific director of ESA, writes and supports the idea that space now constitutes one of the main horizons of mankind: "Confined to the only known inhabited shore in the Universe and for many years to come, confronted with unstoppable limits imposed on him by the laws of nature, energy resources and his own nature, he stands on tiptoe. He cannot condemn himself to the slow suffocation or the ennui that will follow the major ecological make-over, covering the globe with a coat that is clean but restrictive. He reaches out in hope. He looks up at what he knows will take him an age to reach, if indeed he does succeed one day. Soon, great telescopes will show him other shores, and maybe even, the existence of other universes. Man has always enjoyed observing faraway shores, even though they are physically unattainable. He dives with wonder into this ocean of dreams. He becomes an astrophysicist despite himself and perceives even more clearly these chimerical horizons".[60] Based on a perspective with such a markedly mythical dimension, Bonnet's opinion of the space venture, both in terms of past realisations and future projects, is resolutely optimist, and even, as he sometimes admits himself, harsh. So whilst "there are those today who think that the conquest of space will only be recorded in the history of mankind as a disruption of no consequence", Bonnet sides on the contrary with those who think that "the 21st century will be dominated by biotechnology, electronic intelligence, robots serving man and cleaning up the planet, but above all by space colonisation. The first phase that we have just lived through only represents the premises of an irreversible flight of man to new shores in the Universe".[61]

Tied in with the tradition of the pioneer, closely associated with North-America, the myth of the frontier is much related to that of the horizon. However, there is a difference in that the myth of the frontier involves a form of duplicity, or even paradox. In effect, the idea of the frontier is based, on the one hand, on the refusal of man to accept limits on his need to explore, his desire to conquer, thus driving him to cross every new frontier and, on the other hand, it signifies that man will only ever push back this frontier. The myth of the frontier is doubtless one of the many forms adopted by the myth of Sisyphus; it has some of the aspects, but simultaneously presents a dynamism that is lacking in the Greek mythological tale. To attempt to reach a frontier, even when others hide behind it, is not completely absurd and, contrary to the punishment inflicted upon Sisyphus, can have meaning to the person attempting to do this. Even though this person would return to his point of departure, he would never be the same again and would have acquired a unique experience, a new outlook and even what many traditions refer to as a wisdom which, in ancient times at least, does not exclude the idea of know-how. Let us look at the remarks made at the time about the Apollo missions on the future of space venture. In the 1970s, based on the official discourse of NASA, the North American *Britannica* encyclopaedia predicted men going to Mars in 1982, the sending into orbit around the Earth of a manned space station with a reusable shuttle (for conducting biological and metallurgical experiments and dividing the cost of carrying and retrieving test equipment by 200), and another space station orbiting the Moon, also served by a shuttle and facilitating exploration of the Moon's surface, a nuclear-powered shuttle connecting these two orbital stations and finally the launch of satellites into geostationary orbit for television, radio, navigation, air traffic, weather and astronomy. Of these ideas, only the latter has really been achieved and, for almost forty years, no human being has returned to the Moon: did the Apollo missions serve any purpose? Could they be classified as absurd? In terms of the projects of the 1960s and 1970s, whilst certainly ambitious, could the space venture have run out of steam? Even so and whatever the achievements of half a century of space activities, a frontier has clearly been crossed: the Earth has been seen by a human being from another celestial body; it is no longer the only place where mankind could possibly live, despite the Moon and space in general offering only mediocre living conditions. In short, even though Edmond Husserl can write that "the Earth does not move" (see below), anthropocentrism and geocentrism no longer coincide. The myth of the frontier continues to be used, as in the National Space Commission report, appointed in the mid 1980s by Ronald Reagan to prepare the United States space programme for the next 50 years. The title *Pioneering the Space Frontier*, and the ethos of this report are clear: "The settlement of North America and other continents was a prelude to humanity's greater challenge: the space frontier. As we develop new lands of opportunity for ourselves and our descendants, we must carry with us the guarantees expressed in our Bill of Rights: to think, communicate, and live in freedom. We must stimulate individual initiative and free enterprise in space. [. . .] The new space programme we propose for 21st-century America will return tangible benefits [. . .] by opening new worlds on the space Frontier, with vast resources that can free humanity's aspirations from the limitations of our small planet of birth".[62]

In the same vein, and just as mythical, is the theme of utopia. Utopia finds its roots in the work of Thomas More, and illustrations later in Jules Verne's *Extraordinary Voyages*. In the literary genre to which these works belong, science becomes a "fictional subject" which is used not only in travel but also in projects qualified as utopian. These projects are not always easy to understand, varying between pure speculation and hidden truth on the one hand, and between the suggestion for a society and its denunciation on the other hand. Whatever the complexity of this notion, the utopian cause is certainly itself also integral to space. Related to the theme of the quest for the horizon, utopia brings with it the project of establishment in a foreign place (i.e. beyond a frontier, whatever form it may take) and void of all human venture. Why not dream of establishing a perfect society, ridding it of all the flaws which today's Earth inhabitants are forced to put up with? Gerard O'Neill's work and projects provide a surprising illustration of such an attitude: in the space spheres which he suggested building in order to establish human colonies, he imagines creating gardens and introducing birds, insects and butterflies but not cockroaches, mosquitoes or rats: all species that are harmful for humans or which humans find unpleasant will be left on Earth, far from these new celestial paradises![63] Space utopia now appears to be the combination of an expression of the mythical motivations of conquest and flight, horizon and frontier, and the new world. However, let us not forget that, by definition and etymology, utopia is always elsewhere or rather, nowhere.

8.2 Explore in the quest for knowledge, but at what cost?

One of the explorers of the North Pole, Fridtjof Nansen explained: "The history of the human race is a permanent battle of light and darkness. It is not a question of the use to which the knowledge is put: man desires knowledge and when this desire dwindles, he is no longer a man". His compatriot, Roald Amundsen, adopted a harsher approach: "Little brains have only room for thoughts of bread and butter". These two men viewed exploration and quest for knowledge as inseparable. Yet, how many explorers and how many academics have paid with their lives to satisfy their thirst for knowledge, their thirst and that of the human species? What limits need to be set, what price needs to be paid to honour this knowledge?

8.2.1 Towards a greater conscience, in science

"Wisdom entereth not into a malicious mind and knowledge without conscience is but the ruin of the soul", wrote François Rabelais in chapter eight of his work *Pantagruel* (1532). Why not propose and support the symmetrical formulation of this wise sentence? "Conscience without knowledge is but the ruin of the soul": far from creating a blind or blinded apologia of knowledge, it is necessary to recognise the place that human beings and the societies they form must grant to science and to the quest for knowledge in all forms.

The scientific approach seems to go without saying: isn't to know or, more accurately, to seek knowledge, innate to humans? In the past, good scientific conscience has contented itself with claiming that knowledge is innocent and impartial, that the responsibility for its use is down to society, as is its culpability in the event of its misuse. So, if science is thus declared ethically neutral, we must not try to subject it to the constraints of a legislation or an ethics committee .... To my mind, this reasoning is incorrect. In effect, science must not be reduced to knowledge, i.e. a pure activity of reason that could be considered ethically neutral. Scientific activity cannot claim such purity either, as it is mandatorily associated with a social reality and with the powers of an individual, a group or an institution. It is fulfilled by the impetus and using the resources allocated by private or public, national or international organisations. Jacques Blamont demonstrated this in his political history of scientific discovery, entitled, *Le Chiffre et le Songe*: "an institution is the gathering of academics in a given place, their order in a permanent and hierarchical structure, and their practically unlimited financial support, made possible by the wealth of a patron. No science without funds; no funds without political aim. We are familiar with the figure of the patron: he asserts his power through caprice and showcases his glory. This is the fantasy of the rich man. We are also familiar with the enlightened tyrant who prefers the company of philosophers to that of roughneck soldiers. The prince who creates a scientific institution follows a different pipe dream: he wants to establish an empire built on science".[64] Eliminating the idea of a neutrality of science and, moreover, of technological innovation, to raise questions about the type of relation which establishes itself within scientific institutions, with political and economic support, is necessarily related to the ethical questioning to which techno-scientific organisations and space agencies must subject themselves.

Our era is marked by a process of the democratisation of science which definitely encourages us, or even obliges us to adopt and pursue an ethical approach, applied to the field of sciences. Here, the term "democratisation" not only designates demands expressed and efforts made in terms of scientific popularisation, but also (and sometimes even in an influential way) designates claims in favour of a better understanding of the issues and a real say in scientific and technological choices. It would, without doubt, be easier to be able to make a clear distinction between sciences and technologies: yet such a distinction is not easy, or is even impossible, today and we need to accept the existence of certain confusion, effective or ineffective, between these two domains, as demonstrated by movements of opinion with regards to nuclear, genetic engineering and, more recently, nanotechnologies. The perception of the skills of scientists, engineers, political leaders and technicians is not enough today to quash all curiosity or fear among a public that does not have all the facts. Considering this change within western societies is a real challenge for their scientific research and technological development institutions.

8.2.2 Envoys dedicated to science

The study of human beings and the possibility of conducting experiments on them has long given the philosophical, religious and ethical conscience of human societies food for thought. This concern is expressed today in particular in the field of medicine, through the implementation of ethics committees (within hospital establishments on an international scale) to the study of the foundations and the elements making up bioethics. However, these initial thoughts, directly focused on humans, are part of a more general thinking concerning the ethical issues of experiments of any type conducted on living species. Within the domain of life sciences, space can therefore already rely on ethical thinking defined on Earth, in addition to regulations to which it must conform. However, experiments on living species can also give rise to more general ethical issues.

According to the *Nuremberg Code* (1947), all human experimentation requires the prior informed and voluntary consent of the human subject. On a wider scale, biomedical research is now subject to legal, regulatory and methodological constraints. This type of research must not be perceived as a simple extension of the diagnostic and care activities for which doctors have been trained, but constitutes a new activity which conforms to specific laws and regulations. Several types of question still remain unanswered and can concern experiments conducted in the space domain. This presents at least one specific characteristic, that of relying on a few highly motivated individuals making up a group of elite people, who are often both guinea pigs and experimenters. In this context, it is advisable to raise questions concerning the preservation of the medical secret when dealing with clinical research, the applicability of bioethical laws to a laboratory as specific as that constituted by an orbital station, or even in the sense of respect of anonymity and the free consent of individuals. More generally speaking, it would be worth looking at the issue of free consent or informed consent, in addition to that of the acceptance of risks by astronauts beyond the strict confines of human experiments and, more widely, concerning manned flights. The still very singular and exceptional nature of a trip to space adds a specific element to this acceptance: how can it be taken into consideration?

8.2.3 Animals before humans

Claude Bernard was the first to use the animal model in biology and invented the experimental method in medicine. Man and animals in fact share a number of physiological processes. The space domain is not exempt from this practice: from the beginnings of space adventure, animal took the place before man. At the time of the first space flights, the dog Laika went before the Russian Yuri Gagarin and the chimpanzees Ham and Enos went before the American Alan Shepard. This was to gain a better understanding of any physiological changes and pathological consequences that may affect the pulse, breathing, arterial and venous pressure, under the effect of acceleration, altitude and microgravity. More than 35 species of animal have been sent into space over a period of 50 years.

There are many arguments to support the use of the animal model. Those of an ethical nature primarily consider the moral limits in terms of investigation and exploration on a human being. As current legislation stands, biopsies, functional destructions, the implementation of sensors in the inner structure of the body are only possible with animals. In addition, the potential for experimentation on humans is restricted due to the absence of the possibility of any therapeutics in the event of an incident or accident. According to regulations in force, *in vivo* experiments on animals is itself limited to strict necessity. Any which way, there would seem to be three imperative rules: experimental protocol is compatible with health requirements, the procedure used causes as little trauma as possible, clinical examination is performed on a daily basis.

Scientific arguments more specifically concern experimentation conditions and the population of models and their environment must be as homogenous as possible. Astronauts are themselves from a heterogeneous population (physical and nutritional conditions, psychological condition, level of training, degree of familiarity with the space environment, workload) and it may therefore seem difficult to compare the quality of the batch (if I may so call it) they could form, with that of a laboratory animal population specifically selected for their homogeneity and placed in permanently controlled conditions.

Arguments of an operational or methodological nature are primarily related to the only partial availability of crew members, with respect to experiments, notwithstanding their small numbers; tests cannot be multiplied either before or after the flight. On the other hand, tests can be multiplied with the animal model, including on experimental subjects similar to those which took part in the flight. However, it is important to consider the drawbacks associated with the use of animal models in space: the possible unpleasantness in the event of the inadequate sealing of maintenance modules, for flights on which both humans and animals are aboard; the constraints associated with feeding; the retrieval of urine and faeces; the sensitivity to stress.

There have been several controversies regarding the use of animals in space biology programmes. The usual arguments are apparently no longer adequate in responding to uneasiness or doubt among the general public. At the beginning of 1999, the *Space News* review published an article by A. R. Hogan and N. D. Barnard, promoting a greater respect for our "terrestrial relatives", the expression now used to designate animal species. Both authors refer to the unpleasant aftermath of the BION 11 mission, in 1996, and the death of the monkey Multik, and the surprises of the Neurolab mission, in 1998, involving the death of many of the rats on board. They focus on the regrets of Oleg Gazenko, the former head of Soviet animal testing in space programmes. Several weeks later, Gazenko specified that, without going back on his regrets ("We did not learn enough from the mission to justify the death of the dog Laika"), he considered as nothing less than decisive the contribution made by animal experimentation to the progress of physiological knowledge, both demanded and authorised by the development of space technologies. Space does not therefore escape the paradox often apparent in our western societies: animal testing seems to provoke more doubt, reluctance and opposition than tests conducted on humans! Can the reason for this be solely put down to the fact that humans possess a conscience, unlike our terrestrial companions? Can the debate be reduced to the question that appeared several years ago on a web page: "Is planting a flag on the Moon or on Mars worth the life of a dog or a monkey?" In any event, the use of animals for experiments in space was the subject of intense discussions within ESA. In Spring 1999, an agreement authorising the use of laboratory animals in space was concluded between all partners, excluding Sweden. However, several voices spoke out in favour of the establishment of a committee responsible for drawing up ethical rules pertaining to animal experimentation.

Animals in space are given either the status of guinea pig or substitute. In theory, the status of guinea pig does not require any specific comments: firstly, it is comparable with that of animals used by laboratories on Earth and globally raises the same questions, primarily that of the necessity for their use. However, one of the questions and the only one that may be worth holding on to as being specific to space concerns the possibility of managing the well-being of these animals. It would be difficult to send work inspectors or health inspectors to check the scientific practices applied in space stations and spacecraft, especially for unforeseen and discrete inspections. Incidentally this comment also concerns the application of laws and regulations regarding experiments on humans.

The status of substitute is more specific to space. By this expression, I mean animals which take the place of humans on board space vehicles to test their reliability prior to the boarding of humans. The ex-USSR, the USA and also China have sent animals into space in order to prepare for their future manned missions. It is not strictly speaking a question of "backup" (an expression used moreover in the domain of manned flights), as these are intended to replace a faltering human being and not to precede him. In looking at the photographs of Laika or Ham, it is however obvious that the animal literally takes the place of the cosmonaut or the astronaut. Seeing the helmets, suits and harnesses worn by these animals, how is it possible not to talk about simulacrum? Or better still an emissary? Yet are we ready to accept animals as the emissaries of mankind?

A sheep, a cockerel and a duck...

French history tells us that the first air travellers in the history of aeronautics and astronautics were a sheep, a cockerel and a duck. All three were placed in a wicker cage suspended from one of the machines designed by the Montgolfier brothers. On 19 September 1783, they took off from Versailles Château and, after a 10-min flight, set down in Vaucresson wood. The cockerel received a broken foot during the fairly violent landing. However, this incident did not stop the first human voyage from taking place two months later on 21 November 1783. Pilâtre de Rozier, accompanied by the Marquis d'Arlande, managed a longer flight lasting around 20 min. Unlike the three animals which flew before them, the two men were able to stoke the fire of their aerostat.

It is worth going over the pages in writer-journalist Tom Wolfe's work *The Right Stuff* devoted to how American test pilots viewed their colleagues selected for the Mercury space programme. Charles Elwood "Chuck" Yeager, the first pilot to travel faster than sound who did not belong to the group of the first American astronauts, explained to a group of journalists:

"I've been a pilot all my life, and there won't be any flying to do in Project Mercury.

– *No flying?* – [...]

The thing was, he said, the Mercury system was completely automated. Once they put you in the capsule, that was the last you got to say about the subject.

– *Whuh*!
– Well, said Yeager, a monkey's gonna make the first flight.
– *A monkey?*

The reporters were shocked. It happened to be true that the plans called for sending up chimpanzees in both suborbital and orbital flights, identical to the flights the astronauts would make, before risking the men. But just to say it like that!... Was this national heresy? What the hell was it?"[65]

Yeager talks about the heresy of reducing the pilot to the rank of animal. On the contrary, French philosopher Florence Burgat talks about denial: the staging and photographs of space animals try, conversely, to eliminate the animality of the space dog or chimpanzee, to elevate them to practically human level.[66] Do we have to choose between these two extreme positions? I do not think so and prefer to retain the idea of emissary, of being sent in the place of; while this also means accepting and deciding to sacrifice emissaries. Buzz Aldrin, the American astronaut who took part in the Gemini and Apollo programmes reminded us, in June 1998, that he and his colleagues appreciated "the enormous debt we owe the space chimpanzees". And he added: "Now it is time to repay this debt by giving these veterans the peaceful and permanent retirement they deserve". The astronaut's project at that time was to raise the money required to place these chimpanzees in a Texan park, under the protection of a specialist organisation. I have not heard whether this project has been completed, but it at least demonstrates how it is possible, even before drawing up animal rights (those of space or Earth), to consider the responsibilities of humans towards them.

If the requirement to use animals in space arises again, it will necessarily be to explore new domains and to cross new frontiers. This applies in particular to long-term missions. Which technologies need to be established to successfully complete missions lasting dozens of months, in zero gravity and subject to strong cosmic radiation? In fact, I am currently unaware of any discussions regarding animals going on board prior to the first missions with humans to and on Mars. However, if it were to become or when it does become the case, I believe that we will be in a position to better consider the meaning given to this venture, which cannot find its justification in the free and informed consent of a team of astronauts and in the genius of scientists and engineers alone. In other words, the use of animals, whether they are considered as emissaries or substitutes for humans, necessarily raises a question and requires an ethical decision, which may be dealt with by the space community itself: the decision to send a living species into space is based, I believe, on the scope of skill and responsibility of an entire society and not a group of experts. If the scientific and technological stakes of such an operation are real and often immediate, the debt with regards to the substitutes of mankind binds us humans. This debt calls for the question concerning the finality of space exploration programmes involving human or non-human lives to be raised urgently and precisely.

8.3 Can space exploration do without nuclear energy?

The proverb "Slow and steady wins the race" comes to mind, yet even this requires energy! To cross the Atlantic Ocean on board his caravels, Christopher Columbus did not have the horses of Gengis Kahn to feed, but his ships needed wind to be able to sail. Why should past, current or future space explorers be the exception to this rule? Besides the very limited capacities of chemical propulsion, nature and technology offer two other sources of energy: solar and nuclear.

Although the use of solar sails to pull spaceships is still seen as futuristic, solar radiation is already used to provide heat through thermal absorbers or is converted into electricity by panels covered with solar cells. This electricity is then either used directly or stored in batteries. This technology unfortunately has two drawbacks. Firstly, solar radiation rapidly decreases as we move away from the Sun, proportional to the square of the distance, and secondly, solar radiation (in particular UV) diminishes the performance of cells over time. Let us not forget that on Mars for example, dust may settle on the solar panels of vehicles, thus reducing the effectiveness of generators. When solar energy does not deliver, engineers have the other option of nuclear energy.

8.3.1 American and Russian technology

The Soviet Union placed thirty satellites equipped with nuclear reactors around in orbit at an altitude of 900–1000 km. These will return to Earth in three to six centuries' time when their effects will in fact be noticeably less harmful. For its part, NASA launched a number of interplanetary probes equipped with radioisotopic sources, still under the control of the Department of Energy, after consultation with and with the agreement of the American Congress. The risks occur essentially at the time these devices are launched or when they return to Earth and they differ depending whether the generators are radioisotopic generators or nuclear reactor generators.

RHUs (Radioisotopic Heater Units) are thermal heaters. The technology is based on the use of plutonium 238 tablets, in the form of oxide, which release heat during the radioactive decay process. The thermal power needed, generally a few watts, requires a few dozen grams of fissile material.

RTGs (Radioisotopic Thermoelectrical Generators) use exactly the same technology, but the thermal energy produced is directly transformed into electricity by coupling with a thermoelectric converter. The several hundred watts of electrical power generally required mean that much higher quantities of fissile material have to be used; as much as several dozen kilograms. In both cases, the plutonium-oxide tablets are stored in impact-resistant (in particular crash-resistant) containers. The use of this specific technology, based on military applications, is currently exclusive to the Americans and Russians.

Many missions have used or still use RHUs and RTGs. Among the most famous are the two Voyager probes (missions to Jupiter and beyond in the 80s), the two Viking Mars landers (1975–1976), the four landers used in the 96 Mars mission, which unfortunately fell into the sea in November 1996 following failure of the launcher, the Jovian probe Galileo launched at the start of the 1990s that contained 56 kg of plutonium and which NASA intentionally destroyed on Jupiter to prevent its collision with one of Jupiter's natural "icy" satellites, finally the Cassini-Huygens probe launched in 1997 and arriving around Saturn in 2004 with 32 kg of plutonium. The use of RHUs and RTGs is planned in the near future for a number of Mars landers used in the Mars exploration programme. This is the case of the NASA MSL09 (Mars Science Laboratory) rover, which will be launched in 2011. Larger than its predecessors (Pathfinder, Spirit and Opportunity) at the size of a small car, this spacecraft cannot rely solely on the energy supplied by solar panels which are necessarily limited in size and whose surface may become covered in dust. The use of a radioisotopic generator would therefore seem inevitable and unquestionable for engineers and scientists.

8.3.2 Public concern

However, the same is not true of public opinion. Nobody can now ignore that atmospheric nuclear tests conducted as part of past or recent military applications have released several tonnes of plutonium into the atmosphere which gradually falls back to Earth and gradually enters the ecosystem and the food chain. In 1966, an American plane transporting atomic bombs exploded above Palomares (Spain) during in-flight servicing and fall out contaminated an area estimated then to be 2 km^2. Moreover, the badly controlled return of the Soviet Cosmos 954 satellite, has still not been forgotten. It crashed in Canada's Great North on 24 January 1978, polluting a surface area of 600 km^2. To the present day, Canada has received no compensation for decontamination costs from the (ex-) Soviet Union. It is not surprising therefore that in 1997, the launch of the Cassini probe met with opposition from those who feared not only an accident during its launch but also a trajectographic error during the gravitational assistance manoeuvres made by the probe and during which its trajectory would be close to that of the Earth (fly by operations). Opponents launched the Stop Cassini web site and published the Stop Cassini Newsletter, in order to get the American Congress to postpone the launch and use a different source of energy. They referred to the

This is the REAL Stop Cassini web site -- accept no imitations!

Table of Contents

Top News Item: (as of April, 2001)

No RTGs AND no RHUs in space!

NASA is careful to tell the public it's not using Radioactive power sources (generally known as RTGs) for any planned missions to Mars. (Example: Scott Hubbard, NASA Mars Director, April 10th, 2001, CSPAN: "All the missions up through 2005 are solar-powered...").

But they ARE using Radioactive Heater Units (RHUs). Each one provides a modest amount of heat but uses 2.7 GRAMS of plutonium (mostly Pu 238).

Each one has hardly any containment system at all and contains millions of potentially lethal doses of plutonium!

Cassini:

At this website is the science and philosphy behind why Cassini should have been stopped.

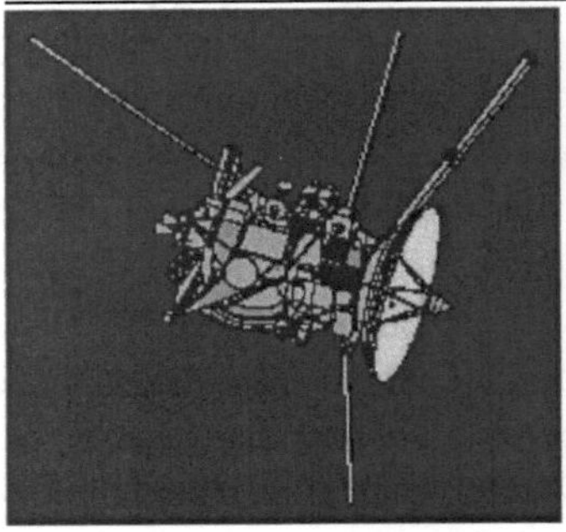

Welcome to perhaps the world's most controversial web site ever!

Newsletters listing

Here is our index of STOP CASSINI newsletters. After issue 253, items were distributed in a loser format. Many of them have been posted HERE.

Fig. 7. *Stop Cassini web site.*

Challenger accident in 1986, and requested that NASA pay more attention to promoting reasons for future hope, rather than provoking fears. The result was that the launch of Cassini was accompanied by exceptional security measures, including the Internet surveillance of information concerning the organisation of any demonstrations. The mission went off without a hiccough.

The use of plutonium carries with it three main risks. The first is its loss or disappearance during mission preparations. This risk is largely contained through checks and monitoring operations. The second risk is directly related to the exposure of operators to radiation during the various handling procedures and integration operations. In fact, in the usual working conditions on probes, the doses received remain well below the theoretically permissible levels due to the implementation of suitable radioprotection measures. The third risk consists in the release of plutonium into the atmosphere, ground or sea, following a launch incident or during uncontrolled return of a probe to Earth: in this case, there is a serious risk of contamination. Plutonium becomes trapped in the organism causing the destruction of tissues and cellular mutations that can degenerate into cancer. In addition, plutonium is an extremely toxic product and between 20 and 50 mg of it if ingested or inhaled is enough to cause death.

8.3.3 Specific precautions already in place

If this technology were no longer to be used due to the very real hazards involved, it would be necessary to launch research programmes on better management of solar energy or the development of other sources of energy without risk, or at least less risk. This would be extremely expensive and, in the short-term, would jeopardise many missions. However, space agencies are on the case and take the matter seriously. They are trying to avoid the use of plutonium as far as is possible as the three recent examples below demonstrate.

The Rosetta mission orbital probe, launched by ESA, uses large solar panels (32 m in size) which are more efficient due to gallium arsenide cells. The probe is capable of going into long hibernation phases in order to save energy on board. In August 2014, this probe will have a rendez-vous with the Churyumov-Geraskimenko comet, at a distance of 5.5 astronomical units from the Sun, i.e. the distance between Jupiter and the Sun (an astronomical unit is equal to the average distance from the Earth to the Sun, i.e. approximately 150 million kilometres). This will be the first time that, at this distance from the Sun, a probe does not use radioisotopic sources. This will be true also of its lander, called Philae, even though the size of its solar panels and the distance from the Sun will not provide the rechargeable batteries with enough power, when it lands on the comet; this will be possible at a later time. Will all devices still work? Will the solar panels become covered in comet outgassing matter or dust? There is no guarantee that the mission will not therefore be compromised. In order to conduct scientific experiments as early as possible, the solution chosen was the use of a stage with high-capacity batteries: over five days, the batteries will procure the energy required to successfully complete these experiments and even to repeat them. This is of course a single operation. If it fails, it will be necessary to wait for the solar panels to be effective enough to recharge the batteries.

Finally, launched in 2011, the NASA Juno probe intended for the exploration of Jupiter, which it will reach in 2015, will be the first Jovian mission to draw all of its energy from the Sun.

8.3.4 Recommendations for the future

In 1992, COPUOS issued recommendations pertaining to the use of radioisotopic sources in the "Principles relevant to the use of nuclear power sources in outer space". These recommendations emphasise how the use of nuclear power sources in space must be based on a thorough assessment of their safety, to include a risk probability analysis, to pay specific attention to the reduction of the risks of accidental exposure of the public to radiation or harmful radioactive materials, and to endeavour to protect individuals, groups and the biosphere against radiological hazards. These recommendations require not only a high level of confidence, but also the assurance of remaining below the acceptability thresholds in all foreseeable circumstances, during and at the end of an exploration mission, and in the event of an accident.

In addition, these recommendations propose that the use of nuclear energy sources in space be primarily limited to space missions that cannot reasonably be carried out using non-nuclear energy sources. If these sources of energy are used in circumterrestrial orbit, it is necessary to prepare for the reorbiting of satellites to a higher orbit at the end of their mission, which contradicts the requirements pertaining to space debris that request a drop in orbit to reduce its stay in space.

Isotopic generators must be protected by a containment system designed and built to withstand heat and aerodynamic forces, during re-entry into the high atmosphere in foreseeable orbital situations, including from highly elliptical orbits or hyperbolic orbits, as the case may be. During impact, the containment system and the physical shape of the radioisotopes must prevent any spread of radioactive materials, so that radioactivity can be completely eliminated from the impact area by the retrieval team.

A State which launches a space object carrying radioisotopic sources on board must inform the States concerned in good time if the object is subject to a failure that may lead to the return of radioactive materials into the Earth's atmosphere. This State is also obliged to help these same States in the event of an accident. If asked by the latter, the launch State must be able to provide the rapid assistance required to locate the impact area of the nuclear energy source on the Earth's surface, to detect the materials re-entering the atmosphere and perform retrieval or clean-up operations to eliminate real or possible effects of damage.

Exploration: this unique human characteristic regards space as a target and a location, clashing with or conversely embracing new space-time perspectives. However, the concerns of the Earth and those of the human race itself linger on. As we gain knowledge and develop our technologies, human identity is enriched and becomes more complex, without it being possible to consider that the question regarding the identity of the visitor itself has been resolved. If it really wants to improve the way in which humans learn, understand and present themselves, space must tackle the question of the "other" and of barbarians.

[56] *The Torch of Life* (1962); *Man Adapting* (1966); *Of Human Diversity* (1974).

[57] Finney, Ben. "Will space change humanity?" *Frontières et conquête spatiale. La philosophie à l'épreuve.* Ed. Jean Schneider & Monique Léger-Orine Monique. Dordrecht/Boston/London: Kluwer Academic Publishers, 1988: 161.

[58] Lesgards, Roger. *Conquête spatiale et démocratie.* Paris: Presses de Sciences Po, 1998: 16.

[59] Ferguson, Eugene S. "Historical Perspectives on Macro-Engineering Projects", *Macro-Engineering and the Infrastructure of Tomorrow.* Ed. Frank P. Davidson, Robert Salked. Boulder (Colorado): Westview Press, 1978: 6–18.

[60] Bonnet, Roger. *Les Horizons chimériques.* Paris: Dunod, 1992: 285.

[61] *Ibid.* 16.

[62] See *Pioneering the Space Frontier. An Exciting Vision of Our Next Fifty Years in Space. The Report of the National Commission on Space.* Toronto/New York/London/Sydney/Auckland: Bantam Books, 1986. http://history.nasa.gov/painerep/begin.html (January 2010).

[63] See O'Neill, Gerard. "The Colonization of Space". *Physic Today.* 27(1994): 32–40; *The High Frontier. Human Colonies in Space.* New York: William Morrow & Company, 1977.

[64] Blamont, Jacques. *Le Chiffre et le Songe. Histoire politique de la découverte.* Paris: Editions Odile Jacob, 1993: 8.

[65] Wolfe, Tom. *The Right Stuff.* New York: Farrar-Straus-Giroux, 1983: 127.

[66] Burgat, Florence. *Animal, mon prochain.* Paris: Editions Odile Jacob, 1997: 149.

CHAPTER 9

9 Invaders

I have already asked the question whether space represents the setting, the opportunity to explore or to conquer. I have already shown how this distinction, as pertinent as it is, was very often caught up with by the reality of our human activity and our propensity to leave tracks behind us, even when we have decided to explore (but not to conquer) the unknown world that surrounds us. In fact, as soon as we leave the surface of our Earth, we pollute outer space physically, chemically or biologically. In return, we cannot exclude or ignore the possibility that we may pollute Earth with elements brought back from other celestial bodies. The question was already asked during the Apollo programme and will become topical again in a few years or dozens of years' time when samples gathered from the surface of Mars or the surface of asteroids are brought back to Earth. In short, any future exploration must account for possible invaders, terrestrial or extraterrestrial.

So then, what should be done? Or rather, what should be done now, as decisions will probably have to be taken as quickly as possible? What principles, what structures do we have? In his book *The Exploration of Outer Space* (1962), Bernard Lovell already raised this concern: is it possible, he asked, to send probes to planets whose biological situation we know little or nothing about? Should we impose constraints, limits to our exploration missions and, beyond that, to conquest? This field of investigation is particularly important and stimulating for space ethics.

9.1 At the frontier of the living

7 August 1996: massive break through for astronomers. The media gets hold of an incredible news story: NASA claims it has proof of the existence of life on Mars! And this proof is allegedly contained in a meteorite weighing 2 kg, unearthed in Antarctica in 1984, whose name was soon the talk of every corner of the globe: ALH84001 (ALH for Allan Hills, the name of the site where it was discovered). David McKay and his team of mineralogists and geochemists identified this ancient 4.5 billion year-old meteorite as originating from the red planet and containing structures 100 nanometres in size and with an elongated shape said to be traces of fossil bacteria. This caused immediate controversy. Opponents of McKay spoke of a possible terrestrial contamination of this meteorite fallen from

the sky or suggested that analyses had been wrongly interpreted. They also remarked that the forms observed by McKay were a hundred times smaller than any other terrestrial bacteria to date. McKay retorted by saying that our knowledge, biological references and conception of life were in the process of changing due to the discovery and study of extremophiles.

9.1.1 What are extremophiles?

Before asking whether ALH84001 indeed contained evidence of ancient living organisms, would it not first be useful to ask ourselves what we currently know about life? The question is delicate and has aroused the curiosity of philosophers and academics for thousands of years. It has given and still gives rise to endless academic skirmishes, to such an extent that it would doubtless be necessary and suffice to say of life what Augustine of Hippo said about time, in his *Confessions*: "What then is time? If no one asks me, I know what it is. If I wish to explain it to him who asks, I do not know". This philosophical question has profound consequences in terms of ethics. This is not the place to bring up the passions that debates about bioethics spark or the confrontations that it may provoke in our societies.

From a scientific point of view, this question is no less inspiring: a life form today is most often and most usually defined as a body capable of reproducing and evolving due to specific metabolic mechanisms and the use of available energy sources (e.g. heat or solar radiation). We already know that viruses belong on the fringes of the living. In effect, these viruses need the DNA of other organisms to self-replicate. The situation is even more complicated with the prion (proteinaceous infectious particle), which is responsible for bovine spongiform encephalopathy (BSE), because it is a pathogenic entity that can be transmitted but is not living; the prion is a singular protein which is extremely difficult to destroy using current sterilisation processes. Now, in addition to finding and identifying reliable biological markers (i.e. molecules or organic compounds characteristic of living organisms), comes the new challenge for biologists of looking for, discovering and studying forms of life in places up to now reported to be totally barren.

The idea of looking further in places where it had been settled life could not survive has partly come from biologists and astronomers who asked which form of life would be likely to be discovered in Mars conditions and, thus, which exploration processes and instruments they would need to detect them. To make headway in their research, researchers decided to examine more carefully than ever before, parts of the globe thought to be devoid of life; and they found life! They called these organisms which live in extreme terrestrial environments, "extremophiles". Extreme environments are those with physical or chemical conditions which exceed the most common terrestrial situations. Studying these extremophiles not only allows us to imagine what life forms on other planets could be like, but also the conditions of survival of terrestrial organisms carried on board probes, intentionally or accidentally, to other planets. Here are several examples, in the form of a veritable parade of the suffix – phile.

If, as their name suggests, anaerobic micro-organisms are capable of living, growing and reproducing in environments deprived of oxygen, barophiles, on the other hand, can withstand astounding exterior pressures: the absolute record appears to be held by spores of the *Holothurians* family which live in submarine depths at a pressure that is 1100 times the atmospheric pressure of the Earth. In terms of an environment's chemical characteristics, nothing seems to ward off life: neither very basic environments (alkaliphiles proliferate at a pH of 10 and some even withstand a pH of 12.5 on a scale that goes up to 14), nor very acidic environments (acidophiles can withstand a low pH; *Picrophilus Oshimae* lives at a zero pH and multiplies at a pH of 1, i.e. in pure acid). Halophile micro-organisms are capable of colonising high-saline environments, whereas xerophiles can survive with very little water and can proliferate directly on rocky surfaces, even in the desert with practically no liquid water available. Pyschrophiles live in frozen waters. The record is currently held by organisms which can multiply at −20°C in the icy waters of the Antarctic, albeit "in slow motion". When it is colder, reproduction apparently stops; but the bacterial spores remain perfectly viable at very low temperatures, e.g. that of liquid nitrogen (−196°C), even −263°C. At the other end of the thermometer, thermophiles colonise very hot environments, with yet more mind-blowing records: 133°C, in the vicinity of submarine volcanic vents. Consequently, it would be impossible to get rid of them using a conventional steriliser, which would only serve as an ideal environment for these extremophiles to reproduce. Radiation is another domain. Some bacteria are capable of repairing their DNA, having been subjected to radiation doses 100 times higher than those used for radio-sterilisation and 3000 times higher than the dose which would kill a human. The conclusion goes without saying: apart from in lava in fusion in volcanic vents or in the depths of our planet, it seems possible to find life practically everywhere.

In short, the world of extremophiles seems completely back-to-front – nonetheless it is also our world, in which we too thrive.

Whilst extremophiles possess extraordinary capacities for adaptation, some bacteria, using the sporulation mechanism, also have equally surprising capacities, with regard to our human biological scales. In a hostile environment, these bacteria can change into spore status, going into hibernation and withstanding external aggressions and resisting time. In 1995, biologists revived bacterial spores found in the digestive tract of a prehistoric bee preserved in amber for more than 25 million years. Having hibernated for hundreds of thousands, even millions or hundreds of millions of years, such micro-organisms are regularly found viable in ice cores taken from the frozen lakes of Siberia or from Lake Vostok in Antarctica. The capacity to sporulate does not only allow these organisms to survive over exceptionally long periods of time, it also allows them to withstand physical or chemical aggression. To kill them, it would be necessary to heat them to over 100°C, subject them to particularly toxic chemical agents and irradiate them with Gamma rays in very high doses. It is because of their remarkable staying power that some strains of bacterial spores are used in sterilisation method qualification protocols (*Bacillus Atropheus, Bacillus Pumilus, Bacillus Stearothermophillus*, etc.). And yet, we are still talking about fellow living beings on Earth.

9.1.2 The revolution of the exobiologists

The discovery of extremophiles would doubtless not have been enough to give exobiology its credentials. To these results from the domain of terrestrial biology we can add the increasing amount of knowledge that we now have about Mars (Mars Global Surveyor, Mars Express, Mars Reconnaissance Observer, Mars Exploration Rovers Spirit and Opportunity, Mars Phoenix, etc.) and of course about other celestial bodies (Saturn, its satellite Titan, and the very recent discoveries made on Enceladus, Jupiter and its satellites Europa, Ganymede and Callisto). Their physical environments and their atmospheres are often studied in order to detect traces of (pre)biotic molecules, or their precursors. Extrasolar astronomy is not lagging behind either: the discovery of the first extrasolar planet in 1995 by Swiss researchers Michel Mayor and Didier Queloz opened new horizons, in addition to new scientific and technological challenges. Today (January 2011), using earth or space telescopes (such as Corot, designed by CNES, ESA and the Brazilian space agency, launched in December 2006; or Kepler, launched by NASA in March 2009), more than 500 planets have been discovered, often leading astronauts to review the planetary system formation models they created based on the solar system. When will we discover a planet sufficiently similar to our own so that we may determine the possibility of life, past or present? In short, as much for terrestrial as extraterrestrial reasons, the exobiology revolution is in full swing.

What exactly is exobiology? Gerald Soffen, one of the main managers of the American Viking missions, explains that exobiology, according to a term drawn up 1960, initially corresponded to the study, essentially chemical and restricted to laboratories, of the emergence and origin of life on Earth. The term was then extended to incorporate the observation of other planets in order to study the possible presence of biotic or pre-biotic life forms. "The science of exobiology, writes Charles Cockell, is the study of the origin, the evolution and the distribution of life in the universe, excluding Earth. One has to admit that, at the start of the 3rd millennium, this discipline has still not had a breakthrough. It is for this reason that, to come up with its hypothesis, exobiology firstly relies on the exploration of our planet's biosphere. By extrapolating our terrestrial biochemical and biological knowledge to conditions that prevail on other celestial bodies, we attempt to make predictions concerning the possible presence of life elsewhere . . . or to explain its absence."[67] Anglo-Saxon researchers make a distinction between exobiology and astrobiology. The latter is more widely interested in the scientific study of the origin, distribution, evolution and future of life in the universe, i.e. astrobiology adopts the exobiological approach, adding to it the study of physico-chemical factors such as gravity, solar and cosmic radiations, magnetic fields and environmental conditions, etc.

French scientists are not troubled by such a distinction. The young French exobiology society (SFE), founded in May 2009, explains on its site that "Exobiology is the study of life in the universe. More precisely, it includes the study of conditions and processes that enabled the emergence of life on our planet and was able to or could enable life to exist elsewhere, the study of the evolution of organic material towards complex structures in the universe, and research that pertains to the distribution of life in all forms it could take, and its evolution".[68] For its part, CNES is interested in the way in which living organisms could be transported by meteorites. The French space agency has conducted several experiments, on board capsules installed in low earth orbit or the international space station: exposure of molecules and micro-organisms to space conditions (Biopan programme); simulation of the atmospheric re-entry of sedimentary meteorites, in order to study their behaviour when they arrive on Earth (Stone experiment); collection of extraterrestrial grains of material to analyse them before they enter the atmosphere (Comet experiment). These experiments revealed that the organo-compounds contained in micrometeorites effectively appear to be protected from solar radiation.

Is exobiology a pointless science? Not to those who are convinced that there is not only life elsewhere, but everywhere: "We are not in search of a Grail", said one of these, Michel Dobrijevic. On the other hand, there are those who believe that only the Earth is able to support the emergence of biological structures as complex as those which populate our planet and which apparently require such specific emergence conditions. In other words, the debate is entered by those who consider that scientists must not allow themselves to become trapped in a geocentric vision of things and those for whom life has to conform to biochemical criteria close to those that we know on Earth. Yet, what exactly are we looking for? And what place, what limits are we willing to give to our imagination? Which set of biological possibilities? This question is not easy to answer. "Planetary exploration, *de visu* then *in situ*, observes Charles Cockell, has [. . .] nearly always shown us to be a little too inclined to unleash our imagination. We have had to backtrack many times and adopt less ambitious hopes concerning present or past life on Mars. And yet, on Earth, there is an extraordinary diversity of living species."[69] The British geomicrobiologist and astrobiologist suggests that we consider life on Earth as a suitable starting point from which to look for life in the rest of the universe. After all, the hypothesis of the extraterrestrial origin of life on our planet dates back a long time.

9.1.3 Primeval soup and panspermia

When and how did life appear on Earth? The question perplexed philosophers, and even Charles Darwin, who, in his autobiography, stated that "the mystery of the beginning of all things is insoluble by us". Even so, there have been endless attempts to provide an answer throughout history and human societies, and this probably since their emergence on the surface of our planet. One of the many suggested hypotheses, spontaneous generation, was widely believed until the 19th century when it was disproved by Louis Pasteur. The famous French biologist managed to prove that mediums previously sterilised then carefully sealed were not conducive to the spontaneous production of living organisms. He also showed that microbial colonies appeared due to contamination of the medium by microbes in the ambient air. So where does life come from? The next suggested response to the still unanswered question was the primeval soup theory put forward for the first time in 1924 by Russian biochemist Alexander Ivanovich Oparin and reiterated in 1927 by English biologist John Burdon Sanderson Haldane. This theory considered that the first elements of life, prebiotic molecules, were synthesized from a "soup" containing water and suitable chemical ingredients. Due to favourable climatic and environmental conditions, it was propounded that these first bricks of life (amino acids, sugars, etc.) combined to form macromolecules. What did these macromolecules look like? Scientists today plead in favour of the form of RNA (ribonucleic acid) which is capable of replicating itself. This would indeed have been a strange world, somewhere between chemistry and biology, a world of pre-cellular forms as described by Francis Crick in the 1960s. The appearance of DNA (deoxyribonucleic acid) may then have resulted in a complexification of living forms and the appearance of the first cells. Before arriving at the current three large families, *Bacteria, Archaea* and *Eucharia,* a common cell, called Last Universal Common Ancestor (LUCA) is thought to have probably existed. The origin of LUCA remains to be determined, as does what might have occurred prior to the primeval soup stage.

The experiment conducted in 1953 by young PhD student, Stanley Miller, is now common knowledge. He attempted to synthesise prebiotic organic molecules in a laboratory by recreating the Earth's environment, as it was thought to have been 3–4 billion years earlier: hot water, an atmosphere comprising hydrogen, methane, ammonia and water vapour, all added to by electrical discharges to simulate early lightning storms. In this way, Miller obtained two important molecules for the synthesis of organic biomolecules, formaldehyde and cyanhydric acid, in addition to four amino acids. However, his experimental soup lacked the many other bricks necessary to create living proteins. What if they had fallen from the sky? This is the hypothesis of panspermia.

The term panspermia is derived from the Greek terms *pan*, which means all, and *sperma*, which means seed. This theory claims that life can begin anywhere in the universe. Greek philosophers in the 5th century AD embraced this theory, believing living beings to have been born of earth and water, sown by the "seeds of ether", a divine, celestial substance floating throughout the cosmos. Nearer our time, in 1865, physician Hermann Richter suggested that the germs of organisms could propagate from planet to planet. The hypothesis was reiterated in 1878 by Hermann von Helmholtz and by Lord Kelvin who supposed that living cells travelled inside meteorites. The modern form of panspermia puts forward the hypothesis according to which primitive Earth would have been sown with a supply of extraterrestrial materials, meteorites and comets, containing life forms or, more probably, organic molecules.[70]

At first glance, the interplanetary environment appears hostile and unsuited to biotic or prebiotic forms that could be found there: absolute vacuum, very high thermal amplitudes according to exposure to the Sun (±150–200°C), and exposure to solar and cosmic radiations. The environmental conditions of planets other than Earth scarcely seem more favourable: frequent absence of protection from solar radiation, absence of liquid water and polar cold on Mars, sulphuric acid rain at 450°C on Venus, atmospheric pressure of 100 bar and −150°C on Jupiter, no atmosphere on the Moon, etc. In short, enough to finish off any survivors of interplanetary travel! And yet, we now know that terrestrial micro-organisms can travel in space, survive on other planets, adapt to extreme conditions and even, if necessary, wait for millions of years for more favourable environmental conditions. In other words, depending on the environment of each celestial body, the chance or risk that living organisms will develop there cannot be considered to be nil.

Two phenomena need to be distinguished here: that of contamination and that of proliferation. In the strict sense of the word, contamination means the supply of foreign forms, on Earth or on another celestial body and proliferation relates to the reproduction and dissemination of these forms. Often, lack of proliferation is perceived as a lack of contamination, understood in the wider sense: is this realistic? Who can be sure that the conditions on a planet, presently hostile to life forms, will not alter to one day allow their proliferation?

So, what type of contamination, or even what proliferation, must we focus on? Which are of concern? I say all, as there is certainly no lack of "means of transport" in space. Before the probes designed and built by humans cross the entire solar system (even modestly), before they disseminate terrestrial organisms there and can one day bring extraterrestrial organisms back to Earth, the flow of celestial bodies between planets has never stopped, in the form of meteorites or dust; the latter are perfectly capable of transporting tens or even hundreds of prebiotic molecules identified in the nebula by scientists. The ice fields of Europa, the geysers of Enceladus, the organic "laboratory" of Titan could harbour such organisms.

In short, the risk of contamination and of dissemination, both for our planet and for the planets we will undertake to explore, must now more than ever be taken seriously.

9.2 Contaminated, contaminant astronauts?

The next time humans set foot upon the Moon or a planet other than our own, should we treat them as samples upon their return to Earth? When asked so bluntly, the question may surprise or even shock. Today, the concern for astronauts most often concerns the composition of teams destined for long-term missions (number of members, *sex ratio*, etc.), the associated psychological drawbacks, the dangers associated with prolonged exposure to cosmic radiation and protection measures to be implemented. This latter challenge appears to be one of the most difficult to deal with to the extent that some researchers question whether there is a risk that the human space epic will be halted either long-term or definitively by cosmic rays.[71] The underlying question is pretty easy to formulate: should mankind travel into space, to Mars and beyond, at all costs?

It may be possible to disassociate public sphere and private sphere. Where is the problem if men and women say they are ready to individually accept the risks associated with travel into space, without possible appeals against public or government institutions, and are ready to implement, finance and complete space exploration programmes? It is also possible to suggest an option other than that advocated by John F. Kennedy in his speech of 25 May 1961; this option was that of bringing crews back to Earth safe and sound. Will the American president's option always be compatible with the dangers encountered by astronauts or the duration of interplanetary travel? Nothing could be less certain, even if there is no reason today to question the challenge posed by Kennedy.

In addition to the risks associated with cosmic radiation, micrometeorites or human factors, there is a risk less commonly referred to which directly concerns the relationship of the individual with society and this is the risk of the biological contamination of astronauts and by astronauts. When we consider the size of the manned modules we quickly understand the impossibility of contemplating a level of decontamination for these structures and their occupants comparable with the level applied to interplanetary probes, at the time of their launch, or the precautions taken during their eventual return to Earth. Based on the experience of the Apollo missions, it is easy to envisage the challenges associated with a manned mission to Mars and to assess how difficult it would be to break the chain of contamination between the Mars environment and the space vessel. How is it possible to prevent astronauts coming into direct contact with Mars dust when they remove their spacesuits in the airlocks? How is it possible to prevent them from bringing this dust back to the modules, then to Earth? It will probably be essential to conduct prior studies during automatic exobiological missions, in order to identify the zones that do not *a priori* present any danger of contamination for astronauts. However, these precautions will not be enough and the contamination of astronauts during a mission to Mars must objectively be considered. Astronauts must actually be considered as potentially contaminated agents and consequently contaminating agents with respect to the extraterrestrial sites they travel to explore and to the Earth to which they will return.

Those trying to answer these questions come up against two major difficulties.

The first difficulty pertains to state of the art and scientific knowledge. For more than half a century, biological research has and continues to grow at an exceptional rate, not only in the field of genetics, but also in terms of the origin and definition of life, as can be seen through the aforementioned recent discoveries in terms of emerging diseases, extremophiles and exobiology. This increase in knowledge, without doubt unprecedented in the history of mankind, is accompanied by the appearance of entire fields to be explored. This tension between the known and the unknown (or to be known) is the very source of scientific research. Without it, researchers would be content to be learned teachers. However, this same tension weakens ventures that provide us with knowledge. Thus, in terms of contamination and planetary protection, COSPAR recommendations are continually being updated as more is found out about planets likely to be explored by our probes. The task is therefore not easy for those who responsible for automatic exploration missions. What can be said therefore of measures to be taken in terms of the risks of contamination of human crews and by human crews, landing on extraterrestrial soil? Engineers and researchers subsequently find themselves faced with the challenge of innovation. Whilst the ideal situation would be to be able to wait to know before acting, more often than not, we are obliged to act in order to gain knowledge. Between knowledge and ignorance, lies experience, the experience of both successes and failures, providing this experience was gathered, analysed and preserved. What value and what priority should be given to the quest for knowledge, to the desire to reduce ignorance, when even the life of the researcher, innovator and explorer is at stake?

The second difficulty pertains to the status of the astronaut. At the slightest risk of contamination, how must they be considered? First of all, without a second's hesitation, like a victim to whom all possible treatment required should be administered; yet also like a contaminating agent to which measures must be applied in order to control, reduce or even eradicate the harmful effects of this contamination for those close to him. In a space context, conventional treatments (vaccination, quarantine) are not easy to implement, even with regard to known contaminating factors. What can be said therefore in a situation where unawareness or even ignorance reign? At the extreme limit, is it possible to consider a definitive quarantine of the men and women who chose to depart as "representatives of mankind"? This kind of decision appears even more dramatic since it is not a question of a condition to which one is involuntarily subjected (e.g. an illness), but a question of free consent, out of any extreme contexts or emergencies, as would be the case for military conflict for example.

Dynamic tension between knowledge and ignorance, status of individuals within the group: the risk of the contamination of a crew of astronauts does not appear to raise any new questions or challenges. It is simply the chance to ask how much risk researchers and explorers, on the one hand, and institutions and societies, on the other hand, would be prepared to accept or to ask others to accept, in the name of acquiring new knowledge and to explore and conquer new domains and territories. Except that the risks faced do not concern only the life of individual astronauts, but potentially that of entire populations. So, how could astronauts, cosmonauts and taikonauts forget the people they represent according to space law and the people whose scientific, technical and financial commitment has allowed them to travel into space? And, even if they only used private funds, they could not totally ignore the risks that they could pass onto the rest of mankind when returning from a mission during which they may have been contaminated.

Without trying to dramatise the question raised here, it has to be taken seriously. In addition, as our knowledge stands, it would seem irresponsible to now consider a manned mission to another planet, a mission during which astronauts may be contaminated, without having previously considered and specified the measures to be adopted to assess these risks and to detect any contamination, the investigation and decision-making procedures in the event of confirmed contamination, finally the measures to be considered to honour these decisions and the means to successfully carry them through, including a possible definitive placement in quarantine.

9.3 Cross risks

During the Apollo 12 mission of November 1969, the lunar module landed not far from the American Surveyor 3 probe which had automatically landed on the Moon on 20 April 1967. Astronaut Pete Conrad was thus able to retrieve the camera from Surveyor 3 and take it back to Earth. After taking the standard microbiological precautions, NASA scientists observed several dozen bacterial spores common to Earth. These were *Streptococcus Mitis* and the scientists even managed to revive them. Conclusion: these bacteria apparently survived the journey between Earth and the Moon and a stay of more than thirty months on the surface of our satellite, subjected to the emptiness of the cosmos and space radiation, an average temperature of −20°C and a lack of food, water and energy. Huddled inside the camera casing, would they have been protected from ultraviolet solar rays during their stay on the Moon? This is highly likely. The validity of this result and of these analyses is now being called into question: some people think that the equipment was accidentally contaminated at the laboratory upon its return to Earth.[72] Even so, the previous pages provide all the proof needed and scientists are now taking drastic steps to prevent or at least reduce the risks of contaminating the celestial bodies they set out to explore. They implemented these measures for the Viking missions by sterilising the two probes to be placed on Mars in July and in September 1976, and also for the French–Russian mission in March 1996, which sadly failed.

The case of the American Genesis probe is also worth mentioning. On 8 August 2001, the Genesis probe was launched from Cape Canaveral by a Delta 2 rocket to collect and bring back to Earth a few micrograms of dust carried by solar winds. These particles could provide scientists with information pertaining to the origins of the Sun and its planetary system. To successfully fulfil its mission, the probe remained on one of the two Lagrangian points, one and a half million kilometres from the Earth, for almost two and a half years. It then returned to our planet. Its return was the stuff of a Hollywood blockbuster. Set against the backdrop of the Utah dessert, on 8 September 2004, at midday, two helicopters were set to capture the probe as it descended under a parachute, reminiscent of the Cold War when

American pilots retrieved films taken by their spy satellites in full flight. Unfortunately, the Genesis mission was deprived of its "happy ever after": its two parachutes failed to open and the probe exploded into a cloud of earth and metallic particles. There had been nothing to suggest that this would occur, until then the mission had run as expected. The NASA-led enquiry identified the cause of the accident. Lockheed Martin, the manufacturer responsible for producing the machine near Denver, had failed to perform tests which were initially scheduled and which would have revealed that the accelerometers had been fitted the wrong way round during probe assembly. The accelerometers did not therefore detect the deceleration experienced by the probe when it re-entered the atmosphere and subsequently failed to trigger the opening of the parachutes. An accident of this type must not be repeated during the Mars sample returns. Readers of Michael Crichton's fiction *Andromeda Strain* (1969) can imagine the possible consequences of an extraterrestrial contamination by a living organism as pathogenic as it is unknown. Luckily, another probe successfully completed the same type of mission and landing two years later. At the end of its seven-year space trip, Stardust returned without incident to Earth on 15 January 2006, bringing with it the equivalent of one teaspoon of cosmic dust taken from the Wild 2 comet near Jupiter. Since the Apollo 17 mission in 1972, this was the first time that extraterrestrial material had been brought to Earth "safely" (J. F. Kennedy).

9.3.1 Highly motivated scientists

The experience of the Apollo programme, the mishaps of the Genesis probe, the success of Stardust and, more recently, of the Japanese probe Hayabusa, and current knowledge in terms of exobiology and extremophiles led scientists, engineers and authorities to take the issue of contamination and, consequently, planetary protection, seriously.

According to Article IX of the Outer Space Treaty: "States Parties to the Treaty shall pursue studies of outer space, including the Moon and other celestial bodies, and conduct exploration of them so as to avoid their harmful contamination and also adverse changes in the environment of the Earth resulting from the introduction of extraterrestrial matter, and where necessary, shall adopt appropriate measures for this purpose". At the request of the United Nations and via COPUOS, the COSPAR issued planetary protection recommendations. It adopted the following policy in 1984: "Although the existence of life elsewhere in the solar system may be unlikely, the conduct of scientific investigations of possible extraterrestrial life forms, precursors, and remnants must not be jeopardized. In addition, the Earth must be protected from the potential hazard posed by extraterrestrial matter carried by a spacecraft returning from another planet. Therefore, for certain space mission/target planet combinations, controls on contamination shall be imposed, in accordance with issuances implementing this policy".[73]

The level of recommendations issued by COSPAR considers two factors: the type of mission and the target planet. The likelihood of contaminating a planet is not the same if the mission involves a spacecraft which only passes near it, an orbital probe which will rotate around it or a lander which enters into direct contact with its environment. The risk also depends on the celestial body itself, the conditions imposed by its environment on an organism from Earth. Recommendations thus relate to crash probabilities to be respected, the levels of decontamination to be reached (less than 0.03 bacterial spores per m^2 for the most sensitive situations, such as Europa, one of the 63 Jupiter satellites, or Enceladus, a Saturn satellite). COSPAR recommends that the system used to bring samples back to Earth be firstly placed on an orbit that does not impact the Earth. The container must be perfectly hermetically sealed and be able to withstand a crash. The first sample analyses must be conducted in high-security laboratories.

COSPAR clearly leaves the matters of motivation ("issuances implementing this policy") and responsibility ("the Earth must be protected from the potential hazard posed by extraterrestrial matter carried by a spacecraft returning from another planet . . . ") open to questioning. What do these terms and conditions entail? Its recommendations concerning the protection of Earth are in line with the Outer Space Treaty: our planet must be protected against all hazards, without any restriction. The recommendations which concern the protection of other celestial bodies are based on a more restrictive vision: the contamination of a body in the solar system other than the Earth is deemed detrimental (or harmful), if it compromises the future of scientific experiments. In other words, COSPAR raises the criterion of scientific interest as a topic for debate. It distinguishes celestial bodies according to their significance in the understanding of the chemical evolution process and/or of the origin of life and asks what the chances would be of terrestrial contamination harming future exobiology experiments.

Although COSPAR appears to respect the integrity of celestial bodies explored, its motivation merits specific attention. What can be said of its restriction to a solely scientific purpose? Must the UN keep to the single motivation of the COSPAR or add others? Which? Article IX of the Outer Space Treaty makes provisions that States will conduct the study of outer space, including the Moon and other celestial bodies, and will continue exploration in a way to prevent the detrimental effects of their contamination. Article VII of the 1979 Moon

Agreement is more specific and seems to take things a step further by recommending that States protect not only the interests of science, but more generally extraterrestrial environments, whatever their theoretical scientific interest. Whilst the international treaties prevail with regards to COSPAR resolutions, this difference between the Outer Space Treaty and the Moon Agreement nevertheless clearly raises the dual question of the motives behind space exploration and reasons to take risks.

9.3.2 Measures, but also questions

Since the beginning of the space conquest, scientists and engineers have been continually developing means to assess the risks of contamination, of the Earth and other celestial bodies, on the one hand, and techniques to reduce or control these risks, on the other hand. Let us start with the latter.

The various techniques to decontaminate vehicles sent into space are now well known and successfully implemented by space agencies. These varied techniques include sterilisation, cleaning with alcohol, the use of sporicides or agents designed to eliminate organic pollutants, assembly in monitored clean rooms, etc. At each stage, microbiological readings are used to check the effectiveness of procedures and the contamination status. These procedures are performed in specially equipped clean rooms, by duly equipped personnel. However, the thresholds recommended by COSPAR are not easy to reach. For example, for all Mars landers, the first level is set at 300 bacterial spores per m^2, which corresponds to a maximum of 300,000 bacterial spores per lander. This level only concerns internal or external surfaces exposed to the Mars environment. The current terrestrial environment presents on average 100,000 bacterial spores per m^2. This means that the procedure performed to reduce this "population" by a factor of 300 is not easy to carry out or to achieve. Even so, to succeed, theoretically ensures a general protection of the Mars environment from a risk of global contamination which could disturb current and future biological experiments.

It should also be noted how well NASA engineers managed the end of the Galileo probe's life (the probe went to observe Jupiter and the Jovian system). After discovering the Europa ice crust, the engineers decided to quickly move the probe to Jupiter in 2003, in order to avoid its crash on Europa. The same issue is currently being studied for Cassini in order to prevent it from crashing on Enceladus. Finally, the Juno mission, scheduled to leave for Jupiter in 2011, will involve the deorbiting of the probe at the end of the mission.

Bringing back samples from outer space poses other challenges. In terms of space vehicles, the external surfaces of equipment brought back to Earth must not come into contact with the environment of another solar system body, if the latter has been scientifically deemed at risk. The samples must be kept in a hermetically sealed container, itself protected by several envelopes, capable of withstanding a crash on Earth and only injected in the Earth's atmosphere once its airtightness has been checked. The other systems, as I have already indicated, shall be placed on trajectories that do not impact the Earth. On board, one or more methods of sterilisation may be applied, if their effectiveness has been proven. When the samples arrive back on Earth, the current rules issued by the World Health Organisation for the carriage and analysis of infectious materials must be applied. It is vital that the samples are initially placed in quarantine, with very strict personal and environmental protection. They must be kept in quarantine for a sufficient duration, to successfully perform biological analyses intended to assess the biological contamination risks. Biological high-security laboratories, BSL-4 (BioSafety Level) type, offer such conditions. The difficulty of these initial analyses has already been highlighted: how is it possible to research and study living forms (or non-living forms such as prions) that are unknown or which present risks that are impossible to anticipate for the Earth's biosphere? It is of course possible to solve the issue by purely and simply getting rid of any possible intruders, by applying an indiscriminate sterilisation technique. This involves the use of very high temperatures and very high levels of radiation, in order to destroy all organic molecules, or at least, to incinerate all biological materials. However, what would be the scientific interest in samples that have been subjected to such a treatment upon being brought back from space?

Assessing the risks of contamination when scientists only have a partial knowledge of the celestial bodies they are exploring, of the life they are looking for, is a real challenge. For example, how is it possible to sterilise a space vehicle, without accounting for the decomposition properties of the planetary or inter-planetary environment in which space vehicles and instruments find themselves? It is based on this predictable and natural sterilisation threshold that artificial sterilisation operations shall be performed on Earth, before departure of the mission, in order to achieve the maximum target and permitted contamination level. How can the danger that a sample from outer space could constitute for the Earth be assessed, without having an idea of the presence of living organisms, of their capacities to withstand cosmic travel, then to survive in the Earth's environment and to multiply, ending up by harming living terrestrial forms or even being pathogenic, and withstanding our sterilisation or destruction techniques? In theory, performing sterilisation without validation and indiscriminately would constitute a contradiction. On the one hand, we do not necessarily know the effectiveness and, on the other hand, it would be detrimental to the physico-chemical properties of samples, making them difficult to use for scientific purposes. In short, it would clearly appear that the risks to be taken are level with the knowledge that these missions must allow scientists to acquire.

False positives, false negatives: to be avoided!

Imagine the exploration of a planet in outer space. Imagine that the area from which samples are taken has been contaminated by the lander, that the lander itself was not adequately decontaminated, or that the instruments used to look for life or signs of life are not themselves biologically clean and sterile. In this case, they risk detecting signs of life, coming from Earth: this is known as a "false positive". Beyond scientific error, a false positive may consequently unnecessarily intensify (and necessarily in the absence of other data) the planetary protection measures of future missions to the planet where the analyses in question were performed. While it is sometimes possible to identify the origin of molecules or atoms, it remains vital to prevent this type of error, as far as is possible, in particular by controlling the biological and organic cleanliness of landers and their instruments.

Let us now imagine that the samples collected on an outer space planet are brought back to Earth and that, during atmospheric re-entry, poor thermal insulation causes them to heat up and sterilises them. When laboratory analyses are performed, no biological trace will be detected and the planet in question will be falsely declared free of biological risk, e.g. with regards to a manned mission. The error can also be made during the handling of extraterrestrial samples at laboratories on Earth. A fraction of the sample is dedicated to the research for signs of life or biological hazard. If stored, handled or analysed inappropriately, this fraction may no longer present any sign of life or biological hazard during analysis. The remainder of the sample is then passed on to the scientific community for further analyses, even though its harmlessness is not absolutely guaranteed. This is referred to as a "false negative", more serious than a "false positive" as it can result in less stringent protection measures.

The solutions provided by scientists and engineers to the challenges raised and to the risks incurred by space exploration and exobiology must be assessed and debated – this is on-going. However, they must not mask the ethical issues surrounding this research, i.e. is this research legitimate? All things considered, who will make the decision to start this research and, for example, approve the arrival on Earth of samples from outer space? Thus, during the winter of 2003–2004, the American Space Agency, NASA and the American Association for the Advancement of Science (AAAS) co-organised a seminar concerning the philosophical and ethical consequences of research regarding the origin of life on Earth and the likelihood of extraterrestrial life. These themes are taken very seriously today by the teams preparing the exploration of Mars and concerned with planetary protection, through NASA, ESA and COSPAR initiative.

I am convinced that it is necessary to apply the principle of precaution, i.e. to implement all the resources we currently have available to prevent the contamination and dissemination of terrestrial and extraterrestrial organisms, outside of their native planet. This in no way means abstaining from any venture or applying a principle of immobilism, but rather of doing everything possible to reduce the risks associated with the discovery and encounter with forms of life other than those that we already know about on Earth. At the same time, it is important to ensure, to preserve a possibility of choice, may be even up to the final stages of scheduled missions. In other words, to try to avoid making an all-to-easy and worthless excuse: "It's too late. The harm's been done", even though (the most pessimistic or the most realistic people will not fail to say it) the Earth may be permanently contaminated naturally by celestial bodies which reach its surface and even though Mars may end up being contaminated, if it is not already, by the sending of probes from Earth.

These precautions will not prevent the emergence of movements of opinion which will react against the taking of such risk, for the Earth, or for other planets, to the point that the existence of such missions is radically challenged. The benchmark here is not the protests sparked off by the use of RTGs during the Cassini mission, but issues concerning the protection of species of animal threatened with extinction or, more generally, by biodiversity. Another question, which is as controversial as it is delicate, is that concerning the rights of animals and non-human life forms. Opinions are now being voiced calling for such a right and the application of "biocentric egalitarianism" or "cosmocentric ethics" (Mark Lupisella). The effort is laudable but seems unrealistic. We are all too aware of the situation with our *Universal Declaration of Human Rights* and all declarations of the same type drawn up by individuals or institutions on behalf of mankind: they are all faced with the difficulty of defining a human due to the very varied philosophical, social and cultural traditions of our species. Based on what law and in what way can we define an equivalent declaration for species that are living or presumed to be living, to which we do not belong? Putting ourselves in the place of another will always be a very delicate operation. Therefore, for the time being, and as far as planetary contamination is concerned, which in fact belongs to the class of major risks, the principles of precaution and responsability appear much more necessary and essential to implement, however inadequate they may be.

Against the return of Mars samples

The proposal of the north-American ICAMSR (International Committee Against Mars Sample Return) is only partially summed up by its title. As specified in its charter (www.icamsr.org/charter.html), ICAMSR's primary endeavour is to raise public awareness of the dangers that a Mars sample return mission could have for the Earth and its inhabitants, whether the sample was directly returned to the Earth's atmosphere or transferred by a space shuttle. Protestors at ICAMSR only see the solution of analysing samples on board the International Space Station as being reasonably safe, notwithstanding experiments *in situ*, during robotised missions. Without any prior studies, they reject the possibility of bringing extraterrestrial samples near our planet. Disguised as a warning, the ICAMSR website homepage reminds us of several of the most tragic epidemics in the history of mankind, including the epidemic that spread among the population of Europe in the mid 14th century, those which wiped out native American peoples during the arrival of the Spanish, etc.

ICAMSR's proposals may indeed have merit, but just how seriously must they or can they be taken and how realistic are they? Can exobiology-related work only be carried out using automatic means or within the confines of an orbital station? And what of the study of the pathogenicity of samples? Or of the capacity of terrestrial organisms to react to attacks from unknown pathogens? A compromise must be found between the excessive concerns of ICAMSR and the peremptory statements of Robert Zubrin, the author of the Mars Direct concept and the book *The Case for Mars. The Plan to Settle the Red Planet and Why We Must* (1996) which claims the planetary protection measures developed by space agencies are impossible, immoral and irrational!

9.4 A vineyard on Mars?

Will the exploration of planets other than Earth be enough for us, once the conditions of travel in space, in particular the speed of travel, have been improved? Will we not be quickly tempted to conquer space, to use its resources, to stay there, for whatever reason: set up scientific laboratories, be tourists, satisfy a thirst to expand, escape the devastated planet Earth? Clearly, no other celestial body in the solar system has the same geomorphological, climatological or biological characteristics (i.e. the same biosphere) as our Earth. What would we need to do therefore to undertake the conquest and the occupation of one of these? The most immediate solution is to adapt humans to these hostile environments. Spacesuits, vehicles, camps and permanent bases: everything must be designed to make human life possible in conditions to which it is not adapted, at least as things stand. What about considering the genetic modification of groups of inhabitants or sending cyborgs, i.e. half-human, half-robot beings; but is this what we or our successors could really consider as "living"?

Another solution, which has already long been considered, is to transform the planet itself to make it favourable to the establishment of life and human societies. With Christopher McKay, I will qualify this installation of a self-sufficient ecosystem on a lifeless planet as "ecopoiesis", i.e. susceptible to undergo a biological change, more or less independent of any human intervention. This NASA researcher remains very cautious in his outlook: the passage from a cold and dry environment to conditions that are milder for forms of life, even simple ones, would require several centuries. The main engine could be carbon dioxide, with the creation of an atmosphere and a greenhouse effect. The possibility of biopoiesis, of the appearance of living beings that are more sophisticated and more complex, would be subject to any organic reserves already present and to evolution processes. The term "terraforming" refers more directly to the creation of a terrestrial type biosphere. This would involve making a planet inhabitable by members of the human species, which would in this case require controlled human interventions.

I will not attempt to assess the feasibility of such projects, neither from a scientific and technological point of view nor from an economical and social point of view. I will not attempt to propound any agenda either. Be that as it may and because this type of project belongs without doubt to Ferguson's MEPs, it gives rise to questions such as: is it for humanity to decide to influence the status and future of a planet to such an extent, to modify certain geophysical and climatic characteristics or to (re)introduce living forms? If so, by what right, at what cost and under what conditions? Finally, why would a human society risk such a venture? Through necessity (e.g. to ensure the survival of future generations), in the name of a law governing the planets and species they would provide a habitat for, to make a profit? Further questions are asked in the case of terraforming and the introduction of human groups to new territories: what composition must be chosen for such colonies, what organisational and governmental systems should be implemented, what links should be retained with Earth? Many science fiction authors (including Kim Stanley Robinson with his 1990s trilogy: *Red Mars*, *Green Mars* and *Blue Mars*), social science and political science (Jim Dator) researchers have delved into this subject. These questions, which appear to be very theoretical or at least far removed from our reality, conceal others which neither stem from fiction or from utopia, but belong to the most ancient philosophical heritage, e.g. that of the rights which mankind must grant to non-human life forms, that of its relation with terrestrial or extraterrestrial nature.

We know that this relationship has been developing throughout the history of mankind, as people and human societies have associated a culture with their nature. Of course, they have adapted to extremely diverse natural environments, but they have also adapted these to their needs. In the West, the rapid development of technologies in mediaeval times and, particularly, in modern times has definitively altered mankind's relationship with the "natural" (provided that this term remains appropriate) environment. Three centuries after Descartes asked us to "make ourselves the masters and possessors of nature" (*Discours de la méthode*, 1637), in 1958, Martin Heidegger was able to observe a distancing between scientifically and technologically controllable nature with regard to natural nature over mankind's existence, both habitual and historically determined. However, contemporary visions of nature and the relations that mankind maintains with it remain very diversified: here, I shall refer to the relationship proposed by cultural theory,[74] as it can be applied to any type of environment, even extraterrestrial. Its authors illustrate the way in which mankind perceives nature as a ball which can move across a surface of varying degrees of flatness; the ball's equilibrium (or lack of equilibrium) represents nature's reactions to human behaviour.

"Nature capricious" is an unpredictable world ruled by chance: the ball moves randomly on a flat surface. Humanity has no power over nature and must content itself with a fatalistic "oh well" attitude. "Nature perverse/tolerant" has a homeostatic ability: the ball rolls over a concave surface with a convex summit. Thus, there is a considerable margin of tolerance and equilibrium, but if the impact is too strong, the ball passes over the lip and the balance is lost. James Lovelock, with his Gaia hypothesis, supports this understanding of nature which calls for "paying attention". "Nature benign" offers a constant balance: the ball is at the bottom of a deep concave surface; no matter what happens, it always comes back to the bottom. Nothing can really endanger this type of nature, which allows an entirely "laissez-faire" attitude. Lastly, "nature ephemeral" is in a precarious balance, which can be

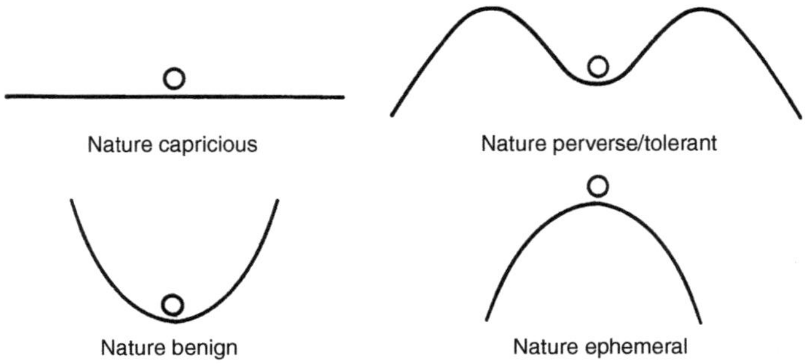

Fig. 8. *The four myths of nature (source: Thompson, Michael et al. Cultural Theory. 1990).*

broken by the slightest impact: the ball is at the top of a convex surface. The attitude corresponding to this vision of nature is "don't touch anything".

It goes without saying that our societies are far from arriving at any consensus with regards to the attitude to adopt with respect to their environments and nature in general. The ideological, historical, cultural and philosophical reasons are so varied that it is difficult to choose between one attitude and another. If this is true of our relationship with the Earth's biosphere, an extraterrestrial (bio)sphere is clearly much more problematic.

Environmentalists often bring up the legend of Noah, builder and captain of a vessel, the Ark, used to save the animal kingdom from the legendary Flood waters. This image is mirrored by others. Another legend exists, similar to that of Noah, which depicts Deucalion, son of Prometheus. Also feeling a flood, Deucalion throws stones over his shoulder to create a new generation. Noah simply opened his arc. This image also reminds us of the image of Spaceship Earth crewed by mankind.

This myth likes to describe Noah as a wise man of ancient times, i.e. a man endowed with an unusual and considerable knowledge, with a clear view of the world, in addition to skills and practical experience. Noah seemed to be a man concerned for the future of the world in which he lived and who, in his own way and in his own time, implemented the principles of liability and precaution. Why not adopt a myth of this kind, next to that of Daedalus and Icarus? Especially his epilogue is very interesting: having become a farmer, Noah the navigator plants a vine to, as the Bible says, create the draught with which he warmed the heart of man and the heady effects of which he was the first person to experience in the history of mankind; it is the relief of a man who had fulfilled his task, who finally found time to rest. Does space venture not itself need a little bit of drunkenness or irrationality, especially if this can promote responsibility? In short, will there be a vineyard on Mars in the future?

[67] Cockell, Charles. "Les grilles du zoo extraterrestre". *La Recherche*, March 2001: 28.
[68] http://www.exobiologie.fr/ (January 2010).
[69] Cockell, Charles. *Op. cit.* 33.
[70] See *Journal of Cosmology*, vol. 7, May 2010, dedicated to the subject: *Panspermia: Transfer of Life Between Stars, Galaxies and Planets* (http://journalofcosmology.com/Contents7.html).
[71] See Parker, Eugene. "Peut-on protéger les voyageurs spatiaux?" *Pour la Science*, 343(2006): 50–56.
[72] See *Earth Microbes on the Moon*, NASA Website: http://science.nasa.gov/newhome/headlines/ast01sep98_1.htm.
[73] *Cospar Planetary Protection Policy* (20 October 2002; Amended 24 March 2005): http://cosparhq.cnes.fr/Scistr/Pppolicy.htm (January 2010).
[74] According to Thompson, Michael, Ellis, Richard & Wildavsky, Aaron. *Cultural Theory*. Boulder-San Francisco-Oxford: Westview Press, 1990: 26–27.

CHAPTER 10

10 The place of humans

In 1610, having received from Galileo a copy of *Starry Messenger*, Johannes Kepler gave his response in his *Conversation with the Starry Messenger:* "There will certainly be no lack of human pioneers when we have mastered the art of flight. Who would have guessed that navigation across the vast ocean is less dangerous and quieter than in the narrow, threatening gulfs of the Adriatic, or the Baltic, or the British straits? Let us create vessels and sails appropriate for the heavenly ether, and there will be plenty of people unafraid of the empty wastes. In the meantime, we shall prepare, for the brave sky-travellers, maps of the celestial bodies – I shall do it for the Moon and you, Galileo, for Jupiter".[75] At the start of the 17th century, following in the footsteps of Copernicus and Giordano Bruno, Galileo, Kepler and Tycho Brahe were not content therefore with smashing the invisible spheres on which their predecessors had comfortably installed stars and planets. Not only did they expel Earth from the centre of the cosmos to set it in motion and launch it into a universe that would otherwise become, if not infinite, at least unlimited, but they had also already imagined that human beings could be set in motion too and would one day leave this planet. Four centuries later, Konstantin Tsiolkowsky outlined the basics of space navigation and astronautics.

Astronomy, the first modern revolution, often qualified as Copernican, was followed by the biological and psychological revolutions, successively triggered by Charles Darwin, Sigmund Freud and their colleagues, fellows and successors. These three revolutions challenged the understanding that humanity has of the world and itself.

10.1 Human nomadism

Although modern sciences and technologies have not (yet?) allowed us to control time (apart from the observation of the universe as it was in a past that goes back billions of years), they have made clear progress in the control of space, not only through the invention of new tools that prolong and make some of our members more efficient, but also, and above all, by procuring the means to travel through space at increasingly fast speeds. Human nomadism at the end of the 20th century and the start of the 21st century has primarily been that of space conquest. In his study *L'espace critique*, Paul Virilio talks about "the emergence of a space-time technology", characterised according to him by the fact that, now, "using instantaneous communication means (satellite, TV, fibre optic cables, telematics, etc.) *arrival supplants departure*", but also that "the direct observation of visible phenomena gives rise to a *tele-observation* in which the observer has no immediate contact with the observed reality".[76] Keeping perspective, the human being who occupies this new space (and it may effectively be better to talk, like Virilio, about space-time?) is as free as he his deprived of movement, as the arrival supplants the departure. It is possible to speak both about an extension of space through the reduction of time and a deletion of geographical distances through current means, mostly electronic, of exchange and interactivity, which is sometimes designated or grouped together under the term "telepresence". Space technologies, which allow man to travel to outer space or to send probes, observation satellites and telecommunication satellites into outer space, doubtless constitute one of the most advanced expressions of this growing control of space by time. It is a true evolution and revolution for mankind.

However, how does this nomadism take shape there? Is it real or virtual? Faced with the increasing gap between imagination and reality, between fiction, simulation and materiality, the objection expressed by philosopher Edmund Husserl in his study entitled *The original ark Earth does not move* is worth consideration. He wrote: "Movement occurs on or in the Earth, away from it or off it. In the primordial shape of this representation, the earth itself does not move and does not rest; only in relation to it are movement and rest given as having their sense of movement and rest".[77] Husserl clearly does not question the Copernican revolution as it stands, but supports the idea of a divorce, too readily accepted, between the sciences, here astronomical, and common sense. Radically structured by time, man is also structured by space, be it only by an orientation, a geometry: a ground and a sky, a right and a left, a front and a back. Husserl continues: "As long as I do not have a representation of a new ground as such, on the basis of which the Earth can have in its coherent and circular orbit the sense as a self-contained body in motion and at rest, and as long as I have not acquired a representation of an exchange of grounds such that both grounds become bodies, to that extent just the Earth itself is really the ground and not a body".[78] The German phenomenologist also mentions the possible existence of other Earths, populated with animals and men. If their existence were proved, would the Earth lose its quality of home, ground, ark of the world and the humanity of its unique character? Not necessarily, affirms Husserl: "The totality of the We, of human beings, of 'animals', is in this sense earthly – and at first is not opposed to the nonearthly. This sense is rooted and has its orientation centre in me and in a narrower We who live with one another".[79] In these words there is no anthropocentric pretension, no exaggerated pride on the part of Husserl and to the benefit of man, but only an observation based on the experience of reality and the correct understanding of phenomena:

man remains at the centre of his own experience of the world, whether at rest or in motion. Jean-Jacques Salomon adopts the following opinion: "Our human horizon remains on Earth, despite scientific and technological progress".[80]

In effect, Husserl's review does not lead us to question the idea raised here of a humanity in motion. It only asks us to remember the place from which this humanity continues to set out on its conquest and the control of space and time. As much in material terms as in symbolic terms, man continues to experience insuperable difficulties in leaving the Earth from which and on which he was born. Humans are Earthlings by nature and, in some ways, in essence, despite the successive revolutions of Copernicus, Darwin and Freud being accompanied by an awareness of the singular (but not necessarily unique), limited and fragile character of the terrestrial biosphere and the potentially harmful, even catastrophic consequences that human activities can have on it. Human nomads today are also beings more than ever preoccupied, tested should I say, by uncertainty and insecurity.

Human nomadism is also accompanied by a change in the representation that each of us, man or woman, has of others, of our *alter ego*. The processes of telepresence have certainly played a part here. It is not necessarily others who have changed, but above all the knowledge that we have of them, in whatever form. High-resolution remote sensing, satellite positioning and tracking, telecommunication and television broadcasting, globalisation, planetarisation and territorialisation are all processes that intrude into the personal life of others, whether we are aware of this or not, whether it is only a means for reciprocal observation or to signal our presence to others. The "future generations" about whom the western conscience is readily concerned, under the persistent influence of Judaeo-Christian traditions, now appear to be far bigger than past generations, and our responsibility to them is increasingly important. Jean-Jacques Salomon is therefore right to ask: "space adventure contributes to the process of internationalisation, but at what moment do the differences and inequalities become blurred? All in all, the odyssey of ethics is not simply about travel into space, and the odyssey of space cannot answer all the questions raised by the scientific and technological transformations we are witnessing. What world have we entered into, not only to promote space adventure, and what place will humanism continue to hold in this world? This is the question we should be asking".[81]

10.2 Towards space humanism

It is probably too soon to truly speak of a space humanism. However, there is scarcely any doubt that, following the example of life sciences, space technologies and sciences raise questions about the necessity of providing new fundaments which can be referred to as humanism.

10.2.1 Being human is about choice

If by humanism, it is common today to mean a thought that essentially attempts to make humans more or truly human, to assist with all that can enhance man in nature and in history, I also like what Jacques Maritain and Max Scheler successively said: shape the world into man and expand man to fill the world. Is this dual movement of concentration and expansion, this sort of breathing, not precisely what space, astronautical and astronomical venture has been offering us for several decades? So, humanism is not a heightened self-importance, a pretentious satisfaction of the human being, but rather a greatness that it acquires and implements, when it seeks to develop all virtualities that are its own, the creative forces of its knowledge, its reason and its genius, within a civilisation and a culture. Understood thus, humanism cannot nor must not hope to one day have a definitive and accomplished form: to its end, in a continuous search and awareness of itself, the human species shall implement its own capacities of adaptation, without losing itself in nostalgia and cautious conservatism, or choosing an ill-considered path ahead. Its capacities of adaptation exist. They surprised a thinker like René Dubos, when he wrote, in *L'Homme et l'adaptation au milieu (Man Adapting)*: "The frightful threat posed by adaptability when the concept is applied to human beings in a purely biological context is that it implies so often a passive acceptance of conditions which really are not desirable for mankind. The lowest common denominators of existence tend to become the accepted criteria, merely for the sake of grey and anonymous peace or tranquility. The ideal environment tends to become one in which man is physically comfortable, but progressively forgets the values that constitute the unique qualities of human life".[82] Dubos does not specify what he means by "the most essential values of human life", which leaves the field open for his critics and his supporters. Does he advocate an integral relativism? That would give priority to movement, to the detriment of a definition, even imperfect, of humans: a little like in the poetry of American poet Stephen Vincent Benet, quoted by Dubos himself "We don't know where we are going but we are on our way" (*Western Star*). However, what could appear to be praise for continuous movement, at the risk of instability and indecision, must be moderated by another Dubos statement: "Being human is primarily about having choice". In effect, for all humans, for all societies or the species to which they belong, all movement, of any kind, is subject to possibilities and drawbacks. These possibilities and drawbacks belong both to space and time and involve major cycles of matter and living forms in addition to vast human peregrinations, with or without return, extraordinary biological and psychological capacities of mankind and the riches and faults of its cultures. Within the possibilities and constraints of this "*jeu des possibles* (game of possibilities)" (François Jacob), every human being can choose and everyone has the freedom to choose. This is their privilege, greatness but also their responsibility and which, even though they are subject to changes and random events of existence and history, no less form the basis of a humanism and of an ethic that does not ignore reality and projects of scientific and technological ventures, but rather seek to steer them, adapt them, make them useful for all humans.

In the West, this philosophical task owes much to Emmanuel Kant, for whom ethics consisted in the free acceptance of an obligation that reason represents to free will as necessary. The idea emerges from a moral law that theoretically draws its sources from pure reason itself and which expresses itself categorically, i.e. unconditional imperatives. Still according to Kant, these imperatives can be understood in the form of precepts of which one of the main characteristics (which forces an opinion in the context of space ethics) is the universality of their field of application, in the name of the moral law and dignity of mankind: "Act so that you can will that the maxim of your action be made the principle of a universal law: Act in such a way that you always treat humanity [yours or another person's] never merely as a means but always at the same time as an end-in-itself". German philosopher Hans Jonas based his opinion on these Kantian imperatives to extend universality in space and time, in terms of responsibility with regards to future generations and all the living. In his *Imperative of Responsibility*, he writes: "Act so that the effects of your action are compatible with the permanence of genuine human life. [. . .] Act so that the effects of your action are not destructive of the future possibility of such life [. . .] Do not compromise the conditions for an indefinite continuation of humanity on earth".[83]

Basing his moral philosophy on categorical imperatives, Kant criticises the subjectivism of Hume who links moral judgements to approval alone, thus inciting emotivism or even pragmatism. The Kantian system also opposes the utilitarianism of Jeremy Bentham or of John Stuart Mill, who have strong influence in Great Britain and the United States. In the eyes of utilitarians, only the examination of consequences enables actions to be assessed or theories to be decided between, independently of any moral judgement concerning the nature of means used and therefore in coherence with the adage according to which "the end justifies the means". That which promotes the greatest happiness of the greatest number of individuals is moral; yet what does happiness really mean. Should we adopt a hedonistic approach and speak in terms of pleasure and absence of pain (Bentham)? On the contrary, do the notions of ideal, quest for perfection of freedom need to be added (Mill)? How can the doubtless inevitable conflict between the sphere of private interest and egotism and that of the universalist altruism be dealt with? Is it not in fact better to give priority to freedom and to well-being, as suggested by philosopher John Rawls?[84]

I do not wish to enter into a secular debate that is more to do with moral philosophy. I would only emphasise that to my mind, the two obstacles to humanism are, on the one hand, the widely spread reference to utilitarianism and to individualism, fruits of a triumphant economic liberalism and, on the other hand, moral relativism (e.g. that which sets out that the rights of man are not intangible and change from country to country and from era to era). It is not a question of judging but rather of observing in the long, medium and maybe even in the short term, that mankind which now experiences space has nothing to gain by blindly following either of these routes. On the contrary. It is better to choose perspectives opened by the principle of precaution and call for caution which, far from disqualifying the intelligence and creativity called for by modern-day men and women, rather map out the frontiers of space and our game of possibilities and judge the possible before asking about permission.

Let us not give these frontiers and this space a purely symbolic value; remember that astronautics, after aeronautics, has successfully rendered the third geometrical dimension accessible to mankind. Whether a question of scanning the enormity of the universe, exploring other planets in the solar system using interposed probes, establishing human colonies beyond Earth or even imagining the more active exploitation of an Earth neighbourhood, in all cases, the pre-copernican closed world has definitively disintegrated, leaving us to deal with a universe whose limits are unknown to us. Moreover, and this is one of the ethical issues of space and space humanism, if space has helped to open up cosmic frontiers, it has also shown how real those of our planet are: the finiteness of the world was replaced by that of the Earth and thus that of its resources. Mankind can no longer shy away from the question of the allocation of resources, now that it knows that its terrestrial expansion is finite and that access to the resources of space is still largely limited. This new approach of the relationship between finite and infinite (which considerably diminishes the relevance of the conventional relationship between human microcosm and macrocosm) must be considered by space humanism. Past thoughts about the concepts of the common heritage of mankind, internationalisation, telepresence and even invaders, simultaneously provide forewords and give the question of otherness the status of pillar of this new humanism.

10.2.2 Otherness, a question that remains open

I like to say that this question follows our humanity about like a shadow. "What is it?" the Cro-Magnon man asked when discovering that the cave next door or opposite was also occupied by an individual that looked like him. "Do they exist?", now ask those who scan the stars by looking out for signals issued by extraterrestrial civilisations or who look at every detail on the photos of planets and celestial bodies that are not too hostile to find signs of life, past or present. The history of the other in humanity is primarily that of an *alter* (meaning different in Latin) become, an (almost) similar, an *alter ego*, i.e. another me. From the first clans to the UN and the European Union, from tribal alliances to the *Universal Declaration of Human Rights* and to the West–East disarmament treaties, the human species has always come up against the effort, the challenge of forming a vast group of *alter ego*, in which each individual possesses, defends or claims the same rights and the same needs. No doubt, the assessment of history and daily goings on offer terrible counter-examples, formidable oppositions to such a process; no doubt, this vision arises from the profession of faith the implementation of which in the domain of education remains a real challenge. However, the movement appears no less real or persistent.

In fact, the human imagination has long populated the cosmos with others forms of existence: gods and goddesses of Olympia, angels of Paradise, monsters threatening the Earth or nice little green men. Their plurisecular encounter, via science fiction but also via religious traditions, has often led to questions about the identity of the human being itself, teaching or reminding us thus that the other is primarily and always at the origin of self. The drafting of messages carried by distant exploration probes and destined for any extraterrestrials that may exist provide many examples. One of the most famous is without doubt the message penned by Carl Sagan and launched in July 1972, at the same time as the Pioneer 10 and 11 heading for Jupiter and beyond the solar system.

This creative inspiration will without doubt be exploited again in the future and I understand the fear that some people have of the idea of discovering that these cosmic others are forever absent, even in their most primitive forms: mankind would lose definitively its imagined and imaginary encounter. However, if tomorrow, or probably at a later date, mankind decided to colonise other planets, or even other galaxies, it would experience without doubt what evolutionists refer to as a biological divergence, at the same time as a cultural divergence: gradually over time due to the geographical distance and the rarefaction or even the impossibility of crossbreeding, Earthlings and settlers, up to now *alter ego*, would become increasingly *alter* and less and less *ego*, in respect to each other. What types of link (social, political, cultural) should they think up to establish between them? This is an interesting topic for future ethical committees and legal experts. Although perhaps surprising to us now, this constitutes the logical follow-on to manned flights.

Whatever the future of ethics may be, it is down to mankind to take every opportunity offered by space to expand its space of possibilities, to enrich its capacities of choice and, in doing so, to increase its freedom, at the same time as its responsibilities. The human being is (or will be) free insofar as, and because, it is moving, in body and mind, in its culture and in its history. And not the reverse.

[75] Quoted in Koestler, Arthur. *The Sleepwalkers: A History of Man's Changing Vision of the Universe.* London: *Pelican*, 1968: 378.
[76] Virilio, Paul. *L'Espace critique*. Paris: Christian Bourgois, 1993: 15, 37.
[77] Quoted in Merleau-Ponty, Maurice. *Husserl at the Limits of Phenomenology*. Evanston (Il.): Northwestern University Press, 2002: 118.
[78] *Ibid.* 122.
[79] *Ibid.* 126.
[80] Quoted in Arnould, Jacques. *La seconde chance d'Icare*. 216.
[81] *Ibidem*.
[82] Dubos, René. *Man Adapting*. New Haven-London: Yale University Press, 1980: 279.
[83] Jonas, Hans. *The Imperative of Responsibility. In Search of an Ethics for the Technological Age*. Chicago & London: University of Chicago Press, 1984: 11.
[84] Rawls, John. *A Theory of Justice*. Cambridge (Massachussets): Belknap Press, 1971.

Conclusion

Odyssey: the title of Homer's poem, which tells us of Ulysses' perilous journey, has now come to describe any tale of travel filled with adventures and, ultimately, any succession of extraordinary events in the life of an individual or group. There is no doubt that this term can be applied without restriction to space venture. Like the Odyssey of the ancient poet, space is abound with gods, heroes and strangers, exotic shores, forbidding places and "Homeland-Earths" (Michel Serres), with brutal combats, Homeric challenges and unexpected alliances. Why should space have to now burden itself with the weight of ethical thinking? Should it not, on the contrary, be free of all constraints to allow humans, all men and all women, to benefit from it and fully embrace it, to find there the chance of success, even if they then return "full of knowledge and reason", to live on good old planet Earth, their homeland, for "the rest of time" (Joachim du Bellay)?

As, even if ethics itself could be perceived as an odyssey, that, observes Paul Ricoeur, of the shift from blind faith (I can do) to actual history (I am doing), is also weighed down by ethics. We are not just talking about the burden, however necessary, of laws, standards and codes of good conduct, but also and first and foremost of the entire burden of thinking: it is heavy, restrictive, hard to think, to commit to a path that forces us to reflect without respite on "how" and "why" we act like we do and make the choices we do. To do this, it is necessary to associate different fields of knowledge of research (science and expertise, history and sociology, philosophy and epistemology, etc.) to obtain the appropriate means required to conduct a critical analysis, to establish an opinion, to make a decision, whether this is individual or collective. That is all cumbersome, I would say, requires time and, above requires the desire to strive to promote dialogue or, to use a mediaeval term, dispute. As a matter of course, opinions and assertions are so often completely different and divergent. Evoking them, gathering them and dealing with them are all part of the ethical approach, necessary prior to being able to give an opinion at all, draw up a standard or enact a law. I am convinced that space, without losing its quality and role of odyssey, must today initiate and pursue this ethical process. Or, to put it another way, ethics can offer space a new chapter in its odyssey, a new frontier to cross.

The pages of this book have described what the first steps and the first adventures could involve: the implementation of regulations and processes aimed at reducing the pollution of circumterrestrial space, the establishment of agreements in order to make data gathered from space as available to as many people as possible, the preparation of exploration missions to other planets and the return of samples, the future of manned flights and space tourism, the surveillance of space and the militarisation of earth orbits, the movement from exploration to use and commercialisation, etc. However, I do not doubt for a moment that other issues will arise in the future, whether or not they pertain to what we are able to predict now. Analyses will be conducted; good and bad decisions will be made. It is impossible to try to say more about it today: space will be what our heirs do in it and with it.

From whom does ethics arise, we could ask ourselves, all things considered. As I wrote, since he has borne the name, *Homo sapiens sapiens* has always practised ethics, most of the time without even realising it. The challenge consists in giving man the means to pursue this task, with increasing efficiency (if such a word can be applied here), wisdom, clear-sightedness, hindsight and ultimately common sense; in other words, in a space of possibilities if not increasingly extensive, at least increasingly known and occupied. Science plays its part without the shadow of a doubt. This must not prevent it subjecting itself to ethical questioning. Moreover, we have witnessed the essential role played by information, popularization and education. This was a key point in the report published by UNESCO in 2000 (which is scarcely surprising given the aims of this international organisation). That said, the idea of a single culture of space is open for discussion: is it possible to imagine a culture that is accepted by all those which make up humanity today and will make up humanity in the future? In which culture(s) will it establish its roots? From the encounter with which traditions will it arise? Nobody knows: this type of event more often than not largely escapes forecasts and control. So much the better.

Rather than contemplating a unique culture of space, or for that matter a unique ethic, I chose to introduce here the idea of space humanism and, in doing so, to try to pre-empt the question. Before questioning (or probably: whilst questioning) whether a space culture will one day exist, it is advisable to think about the human values that space promotes, transforms or demands, about the values that can always come up against the diversity of human cultures, immerse themselves in it and influence. As Jean-Jacques Salomon pointed out to me: "The space odyssey cannot exhaust the questions raised by the scientific and technological transformations which we are experiencing". The world we have entered is bearing the full brunt of these transformations, which may raise issues concerning the ethical foundations and moral *corpus* that have up to now been ours. I believe that the inspiration, the humanist tradition must continue to provide mankind with the reasons for and the means of its existence and its actions. This is clearly not easy "in traditional democratic, yet demobilised and demobilising, structures" (to quote Jean-Jacques Salomon again), but it is perhaps important for us to turn towards sectors of human venture in which a certain idea of the infinite and therefore of the (probable) future continues to reign, in which humans can still let their imagination run free and quench their utopian thirst. So, if the space odyssey merits our interest, if it is worth us embarking on another odyssey, that of ethics, it is not because of the successes, the fears, the risks or failures that space has already provided and will continue to provide in the future. The space odyssey merits our interest not only for the principles on which it is based and will continue to be based (cooperation, concern for common interest, peaceful use, common heritage, province), but also because mankind thus pursues its incessant quest, which has formed and moulded it through its history and which continues to do so: shape the world into man and expand man to fill the world.

List of figures

Chapter 6 A threatening sky

Chapter 7 The Greater Earth

Chapter 8 Exploration

Chapter 9 Invaders

Bibliography

Dealing with the Threat to Earth from Asteroids and Comets. Ed. Ivan Bekey. Paris: International Astronautical Academy, 2009.

Frontières et conquête spatiale. La philosophie à l'épreuve. Published in Jean Schneider & Monique Léger-Orine Monique. Dordrecht/Boston/London: Kluwer Academic Publishers, 1988.

L'Homme dans l'espace. Ed. Alain Esterle. Paris: Presses Universitaires de France, 1993.

Space Use and Ethics. Ed. Wolfgang Bender, Regina Hagen, Martin Kalinowski and Jürgen Scheffran. Volume I: Papers. Münster: Agenda Verlag, 2001.

Text and status of treaties and principles governing the activities of States in the exploration and use of outer space, adopted by the United Nations General Assembly A commemorative edition, published on the occasion of the Third United Nations Conference on the Exploration and Peaceful Uses of Outer Space (UNISPACE III). UN: Vienna, 1999.

Alby, Fernand, Arnould, Jacques and Debus, André. *La pollution spatiale sous surveillance.* Paris: Ellipses, 2007.

Arnould, Jacques. *La seconde chance d'Icare. Pour une éthique de l'Espace.* Paris: Editions du Cerf, 2001.

Arnould, Jacques. *La lune dans le bénitier*, Paris: Editions du Cerf, 2004.

Arnould, Jacques. *La marche à l'étoile*, Paris: Albin Michel, 2006.

Arnould, Jacques. *Qu'allons-nous faire dans ces étoiles? De l'éthique dans la conquête spatiale.* Paris: Bayard, 2009.

Blamont, Jacques. *Le Chiffre et le Songe. Histoire politique de la découverte.* Paris: Editions Odile Jacob, 1993.

Bonnet, Roger M. *Les Horizons chimériques.* Paris: Dunod, 1992.

Crowe, Michael J. *The Extraterrestrial Life Debate, 1750–1900.* Mineola (NY): Dover Publications, 1999.

Debray, Régis. *Vie et mort de l'image. Une histoire du regard en Occident.* Paris: Gallimard, 1992.

Dick, Steven J. *Plurality of Worlds. The Origin of the Extraterrestrial Life Debate from Democritus to Kant.* Cambridge: Cambridge University Press, 1982.

Husserl, Edmund. *La Terre ne se meut pas.* Paris: Editions de Minuit, 1989.

Jonas, Hans. *Le Principe Responsabilité. Une éthique pour la civilisation technologique.* Paris: Flammarion, 1990 (1979).

Lagadec, Patrick. *La Civilisation du risque. Catastrophes technologiques et responsabilité sociale.* Paris: Seuil, 1981.

Lebeau, André. *L'Espace en héritage.* Paris: Editions Odile Jacob, 1986.

Lebeau, André. *L'Espace. Les enjeux et les mythes.* Paris: Hachette, 1998.

Lecourt, Dominique. *Prométhée, Faust, Frankenstein. Fondements imaginaires de l'éthique.* Paris: Synthélabo, 1996.

Lesgards, Roger, *Conquête spatiale et démocratie.* Paris: Presses de Sciences Po, 1998.

Lovell, Bernard. *The Exploration of Outer Space.* New York: Harper & Row, 1962.

Peyrefitte, Léopold. *Droit de l'espace.* Paris, Dalloz: 1993.

Pinkus, Rosa Lynn B., Shuman, Larry J., Hummon, Norman P. and Wolfe, Harvey. *Engineering Ethics. Balancing Cost, Schedule, and Risk Lessons Learned from the Space Shuttle.* Cambridge: Cambridge University Press, 1997.

Pompidou, Alain. *L'Ethique de la politique spatiale.* Paris: UNESCO (COMETS), 2000.

Pons, Walter. *Steht uns der Himmel offen? Entropie-Ektropie-Ethik. Ein Beitrag zur Philosophie des Weltraumzeitalters.* Wiesbaden: Krausskopf Verlag, 1960.

Pop, Virgiliu. *Unreal Estate. The Men who Sold the Moon.* www.virgiliu.com, 2006.

Salomon, Jean-Jacques. "Le risque technologique et l'aventure spatial". *Zénon*. 2 (1997).
Salomon, Jean-Jacques. *Survivre à la science. Une certaine idée du futur*. Paris: Albin Michel, 2000.
Thompson, Michael, Ellis, Richard and Wildavsky, Aaron. *Cultural Theory*. Boulder/San Francisco/Oxford: Westview Press, 1990.
Virilio, Paul. *L'Espace critique*. Paris: Christian Bourgois, 1993.

Author index

Acknowledgements

When, at the end of 1998, Gérard Brachet, then Director General of the French Space Agency, CNES (*Centre national d'études spatiales*), entrusted me with a prospective study of the ethics of space, he had no way of knowing that my research would take me on such an extraordinary intellectual and personal journey. Three years later, I published *La Seconde Chance d'Icare. Pour une éthique de l'Espace* (Cerf, 2001) (*Icarus' Second Chance. Towards an Ethics of Space*), which I presented as the findings of my research, using information gathered from workshops and seminars involving the active participation of a number of my colleagues from CNES and numerous other scientists, philosophers and engineers.

Nearly ten years have gone by, devoted to broadening research channels as yet scarcely explored. Then in July 2009, on a bench at the NASA Ames research centre, Kai-Uwe Schrogl, Director of the European Space Policy Institute (ESPI), offered to incorporate an English version of *La Seconde Chance d'Icare* into the "Studies in Space Policy" collection from ESPI published by Springer. I saw this as a wonderful opportunity to share with a much wider audience the experience accrued by CNES over the last ten years in the up-and-coming domain that is space ethics.

In fact this book is much more than a simple translation, it is more a case here of a work almost entirely rewritten. It expands on the more obvious themes (space debris, planetary contamination, manned flights, etc.) and evokes new ones. It draws on the seminars I attended as part of the International Astronautical Academy (IAA) and the Committee on Space Research (COSPAR), lectures I gave at the International Space University, publications to which I was asked to contribute, and above all the many meetings with those directly involved in the space sector.

As the expression goes, the ideas developed in this book are those of their author and do not necessarily reflect the beliefs of the establishment to which I belong, or the people with whom I have worked and with whom I continue to work. In addition to them, I must thank the successive management at CNES and ESPI for their continued support, in addition to the publishers Cerf, Ellipses and Springer.

About the author

Jacques Arnould, born in 1961, is engineer in agronomy and forestry. He holds a Ph.D. in History of Sciences and a Ph.D. in Theology. He is taking an active interest in the interrelation between sciences, cultures and religions with a particular interest for two sets of themes: the first related to the life sciences and evolution; the second related to space conquest. To the first he devoted several works and publications on the historical and theological aspect. To the second, he is the French Space Agency (CNES) expert in charge of ethical, social and cultural aspect of space activities. He received in 2004 the Labruyère Price from the Académie Française.

MIX
Papier aus verantwortungsvollen Quellen
Paper from responsible sources
FSC® C105338

If you have any concerns about our products,
you can contact us on
ProductSafety@springernature.com

In case Publisher is established outside the EU,
the EU authorized representative is:
Springer Nature Customer Service Center GmbH
Europaplatz 3, 69115 Heidelberg, Germany

Printed by Libri Plureos GmbH
in Hamburg, Germany